The Road to Manufacturing Success

Common Sense Throughput Solutions for Small Business

RICHARD T. LILLY

with Frank O. Smith

SᵗL

St. Lucie Press

Boca Raton • London

New York • Washington, D.C.

Library of Congress Cataloging-in-Publication Data

Lilly, Richard T.
The road to manufacturing success : common sense throughput solutions for small business / by Richard T. Lilly with Frank O. Smith.
p. cm.
ISBN 1-57444-29-6
1. Manufacturing industries—Management. 2. Small business—Management. 3. Success in business. I. Smith, Frank O. II. Title.
HD.9720.5 .L55 2000
658.02′2—dc21

00-012190

Direct all inquiries to CRC Press LLC, 2000 N.W. Corporate Blvd., Boca Raton, Florida 33431.

Visit the CRC Press Web site at www.crcpress.com

St. Lucie Press is an imprint of CRC Press LLC

International Standard Book Number 1-57444-299-6
Library of Congress Card Number 00-012190
Printed in the United States of America 2 3 4 5 6 7 8 9 0
Printed on acid-free paper

Table of Contents

Section III: Common Sense Manufacturing

The Author

Richard T. Lilly, President and CEO of Lilly Software Associates of Hampton, New Hampshire, has founded and presided over three successful software companies since he entered the computer business over 40 years ago.

Mr. Lilly joined IBM in 1960, where he worked alongside Ollie Wight, the pioneer of Materials Requirement Planning (MRP) and, later, Manufacturing Resource Planning (MRPII). Mr. Lilly left IBM to found Software International (SI) in 1975, where as president, he guided the company's growth for seven years. Following the acquisition of SI by General Electric in 1982, Mr. Lilly founded the highly successful ProfitKey International. As president, Lilly grew ProfitKey into a multimillion dollar business.

In 1992, Mr. Lilly left ProfitKey to form Lilly Software Associates, where he created the first concurrent or "finite capacity scheduling" application, which considered both materials and resources when scheduling manufacturing operations. Mr. Lilly obtained a United States Patent on concurrent scheduling in July 1998.

For more information please contact: Lilly Software Associates Inc., 500 Lafayette Road, Hampton, NH 03842; http://www.lillysoftware.com

The Author

Acknowledgments

Writing a book is analogous to contemplating the creation of a company. It begins as an idea. But when you act on it, it soon acquires a life of its own. And as in raising anything from infancy, many people make invaluable contributions to the task of bringing a book to completion. Though I take responsibility for any failings, there are many who gave graciously and freely to the project and to whom I am greatly indebted.

I owe much to all the people in all the many manufacturing companies I have been privileged to work with in my forty years in the software business. These tenacious, creative, and demanding souls have shared their time and concerns and their insights and wisdom, and have added richly to the experience and insights I have gained along the way.

I owe thanks to George Sontag of Heald Machine, who was the first customer of Manufacturing Management Sciences, Inc., my first venture as an independent software developer. Special thanks also goes to Jesse Jones of Associated Machine Company, Inc., who opened my eyes to the unique requirements—and opportunities—of the job-shop world, as well as to Bob Sheldon, the old pro who goaded us into creating the "custom manufacturing" concept in spite of MRP and level-by-level being "king of the hill."

I would like to thank Mary Hubbard of the Worcester Historical Society for assistance in the research of the early history of manufacturing in Worcester, Massachusetts. Thanks, also, to Robert H. Godfrey, director of the IBM Archives, for the generous sharing of his time and the materials in his care.

Numerous individuals familiar with the history of manufacturing and the software industry, both independent of and affiliated with Lilly Software, were gracious with their time in being interviewed on various aspects of the story told here. There include Paul Bacigalupo, Jack Ging, Jim Burlingame, and Walter Goddard, who provided insight into the history of MRP and MRP II, APICS, and the Oliver Wight Companies. Thanks, also, to Loyal Peterman, Jr., Tanya Patrella, and Paul Doyle of Abrasive Technologies, Inc.; Bill Findesien, Ed Hermance, Rob Newton, and Doug Hague of Dearborn Precision Tubular Products, Inc.; and Jeffrey Carignan of Instruments Technology, Inc., for their contributions in telling their stories of success with VISUAL Manufacturing.

Additionally, Skip Casamatta, Bob Davis, Bryan Desjardins, Rich Lagoy, Suzanne Lagoy, Dave Layne, Michael Lilly, Mark Lilly, Frank Maglio, Tony Maurno, Sam Pollard, and Ron Ripley, all of Lilly Software Associates, provided invaluable insight in interviews and valued editorial comments in the reading of the manuscript. Special thanks to Brian Coombes and Scott Rich of Lilly Software for their inexhaustible dedication to bringing the project to completion.

A special thanks to three who were indispensable in the encouragement and support they gave early and long to Frank O. Smith's writing endeavors in the information technology industry. Chris Coleman, Ray Dicasali, and Frank Wingate stand as patron saints to writers and the craft of the well-chosen word in business writing.

Thanks to the 1992 Baxter Lake bunch, who labored with me to build the post-and-beam cabin while we argued scheduling and planning concepts and system requirements for the new VISUAL package: Dave Layne, Rich and Suzanne Lagoy, Mark Lilly, Walter Pigeon, our cook Pat Patterson, plus occasional guest workers Skip Casamatta and Ron Ripley.

I would also like to give thanks to all the employees of Lilly Software, and to all the Lilly Software Associates and their staffs, who have given so much in making the dream of Lilly Software a reality since the company's founding in 1992.

I must also give a heartfelt, if inexpressible thanks to my family, to my sons Michael and Mark Lilly and my daughter Suzanne Lagoy, who not only gave body and soul to building viable, successful software companies, but also labored alongside me in the construction of our home in the Florida Keys, and who unfailingly provided me spiritual and emotional shelter over the years.

And through it all, I owe the greatest gratitude to my wife, Laura, for her constant support over forty-plus years of peaks and valleys.

Richard T. "Dick" Lilly
Hampton, New Hampshire
November 1999

Foreword

As I write this, it is winter in Boston. The weather outside is dreary and cold. It is late in the evening and I am still in the office taking calls from the business press. Unfortunately, a leading software vendor is in a downward spiral and is facing its demise. Clearly, the press smells a front-page story. As I reflect between calls, I think about what a turbulent and tough industry enterprise class software development has become. As an industry it is only some thirty-plus years old but has already seen numerous difficult transitions and a number of companies initially succeed but ultimately fail.

During 1999, the ERP market, as it became known, suffered its first real downturn, as companies locked down their IT shops and invested their monies and time into preparation for the coming of the new millenium—the much feared Y2K problem. As it turned out, the predicted Y2K debacle turned out to be a nonevent in the business world. However, while the world focused on Y2K, the Internet changed the agenda for business. As a result, the manufacturing software industry is facing still another major inflection point. Like a tornado, the Internet has set about rearranging the economy and is reinventing business models. The impact has been felt across the enterprise software industry. Among many others, even industry giants such as SAP have lost their way and along with it their leadership and market share.

In the middle of the industry's trying times, a small software firm led by Dick Lilly continues to grow at nearly 40% annually. Mind you, not by dint of sheer luck or good timing on Dick's part, but by stick-to-itiveness and a belief in the principles of manufacturing and not just software technology.

The Road to Manufacturing Success is about two journeys. The first journey is the story of Dick's own career, which culminated in the VISUAL Manufacturing applications and his firm, Lilly Software. Second and equally important is the historical journey through the technology world and the unfolding of basic principles of manufacturing software. For those who were not a part of this time, Dick's story unfolds the changes, challenges, and market dynamics that shaped an industry that today accounts for over $100B in IT expenditures annually. I know. I was there. Not with Dick physically, but in spirit and work. During the early seventies, I was a manufacturing system designer. I wrestled with many of the same problems, and watched closely as competitors and inventors such as Dick challenged assumptions, created new visions, and competed on a daily basis. Dick's work on the graphic user interface of VISUAL Manufacturing remains for me an icon of creativity and end user insight.

In the beginning, developing manufacturing software was as much art as science. The philosophies of manufacturing business management were only then being codified, often for the very first time. Hampered by the technological limits of early mainframe era computers, Dick had to work in this unexplored world and create some of the very structures of data and processes that continue to be incorporated in most manufacturing software even today. Dick tells the story best as he himself challenged some of his very own models and precepts as computers began to expand the horizons of the possible.

No small undertaking, the pioneering work Dick and a very small group of like thinkers did beginning in the 1960s spawned a huge and important industry, one that easily eclipsed anyone's expectations for how big it would become. And get big it did, the ERP market turned into a $20B industry for the software itself—all based on the original concepts laid down by Dick and his compatriots in those early years. Beyond that, when one considers the hardware, consulting, and other services, the ERP industry explodes into a marketplace that has a total impact that exceeds $100B in sales annually. These dramatic sales figures only underscore the importance of the ERP industry. The sales occur only because, as any businessperson would tell you, there are real savings and competitive advantages in using these systems. Today, billions upon billions of dollars have been saved and countless wealth generated as a result of the very ERP systems Dick Lilly helped nurture. In fact, even the great economist and Chair of the Federal Reserve Bank, Alan Greenspan, has spoken of these technologies as key productivity drivers in the past years' exploding economy.

While continually keeping an eye on the shifts in technology, Dick has focused on the essence of business, helping people make money with their manufacturing business—especially the small enterprises that over the years have become the primary benefactors of the Lilly designs. In Dick's work there is a subtlety to the designs—one of seeing the true essence of a problem and solving it and sticking to his guns despite the trends, power players, and pundits. Needless to say, none of this came easy. The early years of the manufacturing software industry were rife with naysayers, misguided consultants, and an uneducated manufacturing audience that was often swayed by the more compelling speakers, not necessarily the people doing the blocking and tackling in such a tough game. True to his own form, Dick Lilly remained intense, opinionated, and despite numerous setbacks, focused on moving the art of manufacturing software and ultimately the manufacturer forward. One will see in this book that Dick gives credence to Thomas Edison's famous maxim, "Invention is 1% inspiration and 99% perspiration."

That Dick named his software solution VISUAL Manufacturing is not surprising. That is the brilliance of Dick's insight—he saw the needs of the small enterprise as demanding clarity and speed—a total contrast to the rest of the manufacturing application marketplace, which was competing on features and functions that resulted only in increased complexity and escalating prices for the software. Not only rich in deep manufacturing understanding, the software had a user interface that was not only unique, but downright innovative. The innovation continues today, as Dick recently received a patent for an advanced scheduling capability that is arguably the underpinning philosophy of many competitive products in the supply chain software market today.

By itself, *The Road to Manufacturing Success* would make a great history of the ERP industry. To read this book is to gain an insight into the history of the software development industry that would be near impossible without having been there—an inside glimpse to the challenges, politics, and travails that many, if not all, the software pioneers experienced. Several of the younger analysts in the manufacturing software field in which I work—those who entered into the space at a time when the industry was in full blossom—noted that this book was an education; others viewed it as a revelation on how the industry came to its current form. This book is an important and vital document. I recommend that anyone trying to understand the dynamics of the software and computer industries look no further.

David Caruso
Boston, Massachusetts
March 2000

Prologue: The Road to Ocean City

White Plains, New York, is only 247 miles north of Ocean City, Maryland, but it was a distance that took me thirty years to cover. In the journey, I learned more than a few things about business, how to run a business, specifically a software business. I learned a few priceless kernels about human nature as well, my own especially. In the passage, in a sense, I grew and came to maturity, as did an industry that barely existed beyond a faint glimmering in the minds of a few dreamers.

In that time, I built two businesses from the ground up. I walked away from one when it ceased to be fun anymore and was fired by a board controlled by venture capitalists at the other, right as I was becoming convinced the long-awaited potential of technology—that which we, as an industry, had been long promising in glossy marketing campaigns for years—was about to be realized. The irony of the timing of that moment did not escape me.

I did not set out intentionally for Ocean City when I first went to work for International Business Machines in 1960. But serendipity has a curious way of shaping one's course. Memorial Day Weekend, 1992, newly unemployed without a shred of a golden parachute to soften my fall, Ocean City lay out the front windshield of my car like a mythical beacon shining somewhere several hundred miles distant at the end of the stream of traffic I was caught in, rolling en masse toward the shore.

It was a vibrant moment in my life, surreal in its clarity, as such moments typically are. I even had the temerity to be excited. But starting over—more accurately, dreaming of starting over—in your late fifties has hurdles one cannot begin to fathom when starting over in your thirties or forties. Both

I had done before. But with less than $50,000 to my name, I was banking more than just modestly on the hope that I had indeed learned a few things about the business along the way.

It all had to do primarily with the notion of assumptions. The beauty of assumptions, typically, is that we rarely think about them. They go unquestioned, but are immeasurably important to almost everything we do, from getting out of bed in the morning to the rationale for focusing a life's energy toward achieving some singular goal. We assume that with the light of dawn, there is value in surrendering sleep to engage in the activities of the day. We drive to the office feeling safe in the assumption the other guy will stop at the red light rather than barrel through the intersection and take us into oblivion as we enter the intersection under a green light. We work hard at our jobs assuming we'll at least get paid for our efforts, and possibly reap some deeper satisfaction in our tasks.

Thirty days earlier I had been summarily discharged by the board of ProfitKey International, Inc. I had started the company on a shoestring in 1979 and had leveraged it into a successful manufacturing software company in a brief run of years. But ProfitKey, like many other thriving software companies of the early 1980s, had encountered difficult times brought on by rapidly shifting preferences for technology, sparked primarily by the personal computer revolution. Technology began to shift dramatically in the waning years of the decade, as the price/performance of the PC began to heavily erode the entrenched base of the mainframe computer. Prominent mainstay system vendors began to look upon their large installed base of customers as "boat anchors" producing horrific drag on their ability to shift to smaller platforms. Small software startups deploying the latest in technology were the young "Davids" to the behemoth "Goliaths" lumbering along trying to figure out how to support a two-product development strategy. ProfitKey was no exception.

I knew what we needed to do, but was hamstrung by thinning margins and a board looking for greater profits. The marriage of convenience was stressed to the breaking point. The board voted no confidence in my reign and I was, in a manner of speaking, walked to the parking lot, hat in hand, and told to go home.

It was not an easy indignity to endure. It was a stress-filled experience—but I was excited, too. For I knew—*and arguably better than anyone else in the business*—where a motherlode of untapped market potential lay hidden at the very feet of every manufacturing software vendor in the business. It lay hidden behind assumptions; assumptions that had been made 30 years before in the very first design effort of an integrated manufacturing application. I knew this because I had been a member of the small, select team of people who had scoped the design for IBM. And while the technology, the functional breadth of manufacturing applications, and the critical importance of manufacturing competitiveness had all undergone profound transformations in the ensuing 30 years, there had been almost zero tolerance in the industry for systematically questioning the early assumptions upon which manufacturing information technology was based. It was as though what we had done 30 years before was inviolate, set in stone. But now, with the advent of truly business-class PCs and the Windows environment, we had the technology capable of breaking through the resistance by exploiting the weight of end user acceptance. And in this set of circumstances, I was absolutely convinced, lay the golden opportunity I wanted to exploit in starting another software company.

Memorial Day weekend, 1992, I was on my way to Ocean City to meet a man I'd never met before, to make him an offer I hoped he couldn't refuse. I was confident I would at least get a proper hearing. The card in my favor was his son, David Layne, who had been an employee of mine since I'd first hired him as a contract programmer back in Marathon, Florida. He'd been key to the success of ProfitKey. One could even say vital.

At the time I hired him, David knew more about outboard motors and how to read water over a reef than he did about software. But he had a rare blend of innate curiosity and amusement, always wanting to know how, exactly, something worked. He had proven to be not only a natural, but also a truly gifted programmer. It was uncanny how David took to programming when he joined my son Michael and me in the small startup that grew to become ProfitKey. It was as if he could glimpse whole design panoramas from bits of detail he would draw out of me. He referred to it as simply "connecting the dots" after he saw the "theme." I came to trust him implicitly in the foundations he laid in the elegant code he wrote.

My vision, heading south to Ocean City to talk with David's father, was to take my 30 years experience in the software business—having started out before the term software was even associated with what we were doing—coupled with David's genius for programming, and create a manufacturing application system that truly served the critical needs of the customer. I wanted to build on some of the pioneering work Dave and others of our team had done in the early years of ProfitKey. But I wanted to go much further, too. I wanted to design and build a manufacturing control system the way it should be done to support the way the business of managing production is meant to be done. Due to the set of design assumptions we felt compelled to make when we designed the first integrated manufacturing application package 30 years earlier in White Plains, New

York, we had effectively forced the entire American manufacturing base—certainly every company that had adopted manufacturing resource planning (MRP II) applications—to alter the way they managed production to conform with our set of assumptions. The argument could be made—and I believe strenuously—that it had cost American manufacturers billions in lost revenues, legions of unhappy customers, and countless failed businesses.

Ocean City sits hard on the Atlantic, a dot on the map that causes thousands of families from all across the interior mid-Atlantic region to flock en masse every summer, commencing precisely with Memorial Day. That weekend is like a seasonal homing signal. The city lies astride a strip of Barrier Island located between the Delaware and Chesapeake Bays. It has a logic a programmer would love, laid out in neat blocks that run from First Street in the south, down near the Boardwalk, up to 142nd Street, which borders—literally—the Delaware state line. In the off-season, there are about 50,000 souls who enjoy the solitude of the wide Atlantic staring them full in the face. In the summer, the population swells to well over 300,000. Hotels, motels, apartments-by-the-day and -week run up the strip from the Boardwalk all the way to the state line. Besides sleeping, people have to eat. Ocean City may not have quite literally a thousand restaurants, but it certainly seems that way.

Joseph Damiano owned two of them at the time, both named The Olive Tree. Known to everybody as "Buddy," Damiano had moved to Ocean City in the early 1950s to open the first Olive Tree on a site where his sister had been running a nightclub. Good, basic, hearty Italian fare was standard. When I drove down to meet with him, Buddy didn't know a disk drive from a modem, but his Imperial Crab Pizza was without peer.

Damiano is easygoing, with an optimistic outlook on life, an attitude that is priceless in the restaurant business. He and David's mother divorced when David was four. David had grown up with his mother and stepfather in Florida, but he and his father had remained close. Though David insisted we could have negotiated terms to a workable arrangement with his father by telephone, I wanted to meet the man. He had fathered an amazing prodigy. And, after all, I was going to ask him for $30,000, no small figure on his books or mine. That sum would front David's salary for twelve months—what I figured was the bare minimum in time we needed to get our project off the ground. When we sat down together in the television room of his spacious, two-story home, we started from positions that were actually very similar in perspective. He and I are about the same age. But more than age, we both shared a common desire to give his son the opportunity of a lifetime—if we could pull it off. Gut instinct told me that the software market was more than primed for what we had in mind.

Sitting with David and his dad there on the edge of the Atlantic, I knew what I was proposing was the kind of thing that would really get David excited. He'd spent ten years learning the craft of software programming; this was an opportunity to take all that he had learned and set out on an adventure to create something that had never existed before.

I explained in general terms what we wanted to do, how it was dependent on David and I doing it together. I had a vision of how to build a manufacturing software system the way I believed it always should have been done. I saw how to shatter the 30-year conceptual bottleneck that had hamstrung system designers and manufacturing executives alike—limiting the former creatively, and the latter financially, ultimately competitively. I wanted to exploit the newest technology available, which for the first time in 30 years made it possible to deliver on what had long been promised. But I needed Buddy Damiano's son to do it. I needed David's native brilliance to grasp the conceptual, connect the dots, and make it real.

I desperately wanted to connect the dots, so to speak, of my own experience, those that stretched all the way back to White Plains and my first days at IBM. I'd seen much. I had been integrally involved in the development of the software industry, witnessed its inception, and seen it explode into a multibillion-dollar business spinning the wheels of industry around the globe. It was an opportunity I didn't want to miss, even at an age I should have been thinking, perhaps, more about how to entertain myself in retirement. Opportunities like this, however, come along—if you're lucky—only once in a lifetime.

SECTION I: BIRTH OF AN INDUSTRY

1 Front Line

We live with a few familiar ideas. Two or three. In our encounters with worlds and men, we polish, we transform them. It takes ten years to have an idea all one's own—about which one can speak. Naturally, it's a little discouraging.

Albert Camus, **THINK Magazine**, IBM employee publication, April 1960

I went to work for International Business Machines in April 1960, newly recruited in a wave of new talent joining the company. IBM was processing new employees that spring en masse, it seemed, as if assembling ground troops for a land invasion. In a sense, it was. I was in a wave of new hires that swelled the employee roster that year past 100,000 for the first time in the company's history. It was the part of a major push to position the company for dominance in the fledgling commercial computer industry.

Then and always, IBM was a hardware company, long the leading supplier of "office appliance" equipment: typewriters, calculators, sorters, printers, and miscellaneous other mechanical devices that the expanding range of twentieth-century industrialization demanded to keep financial tabs on the growth of corporate empires. Computer technology, greatly advanced as a result of military-funded research and development during World War II, was well entrenched in the plans for the future of a score of major companies looking to fatten their bottom lines and increase shareholder value. By 1960 there was a certain excitement surrounding the notion of the commercialization of computers, an excitement that had made itself felt even in the larger, broader, common culture of the day.

Though the first commercial computer application was a payroll system at GE Appliance Park in Louisville, Kentucky, installed in 1954, Walter Cronkite had popularized the notion of the dawning of a new age two years prior during the 1952 presidential elections. To a national television audience, he christened UNIVAC "this marvelous new thinking machine" when it accurately predicted Eisenhower over Stevenson by a landslide only hours after the polls had opened. (Announcement of the computer's wizardry was actually delayed for several hours, as the rapidity of the accomplishment was so startling there was little initial faith in the projection.) By the time I joined the company, IBM had long since determined it imperative to lead the advance into this new age. As for myself, it was an auspicious time to be a young professional; I had already acquired five years experience in the manufacturing industry, including having worked for Bowl-Mor, a small custom manufacturing company that specialized in the fabrication and installation of bowling pin setting equipment.

Following the Allied victory in World War II, the United States emerged as the world's military, economic, and industrial superpower. Though it had only six percent of the world's population, it produced better than 40 percent of the world's industrial output. True, the Russians had beaten us into space with the launching of Sputnik in 1957, and unemployment had risen to 7.7 percent in 1958—the highest since the Great Depression—but there was a growing sense that the future was ours to claim. John F. Kennedy captured the essence of this in his inaugural address in 1961, challenging the nation as a whole to embrace the "new frontier."

There was more to this than mere hyperbole. Explorer 1, the first successful American satellite, had followed Sputnik into space in 1958, the same year that Pan American Airlines began regular jet airliner service between New York and Paris. In 1959, NASA put Pioneer 4 into space, sliding it past the moon on its way into distant orbit around our sun. That same year, NASA had named the first seven astronauts for space flight.[1] Though we were possessed by lingering insecurities, and world events could still prove capable of producing frightful headlines, it was, indeed, a great time to be both young and American.

As a nation, we entered the 1960s reaping an unprecedented harvest in productivity and standard of living. Our good fortune rested, in part, on "the five pillars" of strength that distinguished us from every other nation on earth. Our domestic market was eight times larger than the closest contender, permitting us much greater economies of scale. We also had been spared the devastation of widespread destruction of our infrastructure during the war, enabling us to move aggressively forward, fine tuning our emerging technological capabilities while the rest of the world was clearing away the rubble. Additionally, we benefited inestimably from the large-scale in-migration of intellectual capital as scholars and scientists from around the globe sought liberties and opportunities unparalleled elsewhere. We had the best-educated, most skilled work force, earning a per capita income that was eight times the world average. And we were unsurpassed in manufacturing prowess.[2]

Growing up during the War and its aftermath, my sights were focused only on wanting to accomplish something of value with my life. What that was, was always rather vague and nebulous in my mind. I was the oldest of four children. My father was a small-town lawyer in the hills of central Massachusetts, a Roman Catholic of strict Jesuit persuasion; my mother a graduate of Simmons College, was singularly a homemaker until my father died.

I was always a good student, receiving high honors in high school. My appointment to attend West Point after graduating from high school was proof in my mind that my heading in life was following a proper course. It had always been my father's ambition to go to the Academy, but he'd been denied entrance for failing the physical. My graduation in 1955 from West Point gave me at least, I believed, a clear leg-up on the old man.

When I joined IBM, the company's sales organization was organized by industry segment. The chief industry sectors were insurance, banking, government, distribution, and manufacturing. I was hired as a manufacturing system engineer. It was my job—and the job of others like me with real world inventory and production control experience—to validate the premise that computers were practical tools for business application, not simply something out of Buck Rogers science-fiction comics.

As a systems engineer, I worked both pre- and post-sales, helping IBM customers first rationalize, then realize value for the princely sums demanded to own a piece of the future. For the expense, if not the pride of novelty in owning one in that day, the first computers had about them the essence of "shrines... rococo cathedrals of refrigerated wires and tubes, costing millions...."[3] They were kept in sanitized quarters on raised daises in remote corporate offices. They were shielded by glass walls, like sacred art—the better to be displayed, but with the assurance they couldn't be touched. They were relegated to the care of a specially trained clergy reporting directly to the chief officer of finance.

My job was all the more challenging, for I was a manufacturing systems engineer, assigned to provide aid to a realm at the far periphery of all corporate kingdoms, one bearing the indelible smudge of the factory floor. As seen from the towers of corporate America at the time, manufacturing was viewed fundamentally as a cost, a necessity perhaps, but something to be squeezed, not valued. We were begrudgingly permitted egress to the shrine that housed the corporate computer only after hours, after finance had been served and gone home for the day. We were admitted in the dead of night, but were to be gone, without leaving fingerprints, before first light.

I was assigned to work out of IBM's Worcester, Massachusetts, regional office. For anyone connected with sales in manufacturing, Massachusetts was a great place to be. Massachusetts was number one in New England and ninth in the nation with industrial revenues in the early 1960s. And Worcester was one of the key manufacturing centers in the state. Worcester has a rich heritage in manufacturing, the history of the city and surrounding area woven seamlessly into the currents that propelled the advent and flourishing of the Industrial Revolution and the making of the American industrial legacy.

The community of Worcester had a spotty history as a colonial upstart. It was settled and abandoned twice before permanency took hold in the early 1700s. For its troubles, however, it was ideally suited for prosperity of the time. It offered an abundance of potential mill courses along the streams that traced their way down numerous commanding hills that encircled the growing settlement. Water was the key power of industry of the day. By 1800, the town boasted the presence of furnaces and forges, tanneries, distilleries, sugar refineries, breweries, and all types of mills, from chocolate to gunpowder and textile mills. All exploited water power in some fashion to support the material needs of the nearly 2,500 people who called Worcester home.[4]

Worcester grew to become a great and prosperous industrial center during the last half of the nineteenth century, as the fruit of invention and ingenuity was spread with the expansion of the national network of rail lines that opened distant markets for its goods. Nourse & Mason of Worcester produced and shipped plows that helped put the tough Midwestern prairie to cultivation. Joseph Glidden shipped barbed wire from Worcester to be used to fence vast expanses of the treeless west for cattle production.[5] In addition, Worcester mills and factories produced wrenches and machine tools, carriages and railroad cars, fire arms and steel, steam engines and boilers, shoes, textiles, paper, women's corsets, pianos, and a plethora of other goods required by a civilized and settled nation.[6]

Interestingly, it was the phenomenal growth and success of American railroads, fueled in large measure by the industrial output of countless industrial centers such as Worcester, that created the first wave of demand for information processing technology in the late 1800s. With the growing complexity of shipment schedules, rate structures, number and kind of rolling stock, and employees in the thousands spread across dozens of states—not to mention hundreds of millions of tons of goods shipped annually—railroad management was hard pressed to keep up with the deluge in numbers generated daily in its course of business. Railroads employed armies of clerks working in ledgers at long tables. Consequently, railroads were first to perceive the general utility of the mechanical tabulating device promoted during the tabulation of the 1890 national census. Using the mechanical computing device developed by Herman Hollerith, work on the 1890 census was completed in just two years—not the standard full ten—netting $5 million in savings, and essentially providing proof-positive of the value of the new punch-card technology. By 1920, mechanical tabulators were deployed throughout American railroading and finding receptive markets in insurance and industry, as well.

Hollerith exploited the same conceptual design of the Babbage Analytical Engine, developed—but never patented—by Charles Babbage fifty years before. Both men borrowed from the design of the automated loom that was invented earlier by Joseph Marie Jacquard, a Frenchman, who devised the method of using punched cards to control the weaving of delicate, intricate patterns in bolts of cloth.

Hollerith's company, the Tabulating Machine Company, merged in 1911 with two other "high-tech" firms of the day to become the Computing-Tabulating-Recording Company (C-T-R). The company hired former NCR executive Thomas J. Watson as general manager in May 1914. It was Watson who guided C-T-R through its metamorphosis into International Business Machines, keeping the company abreast of the field through a three-pronged strategy of relentless new product innovations, a popular equipment rental program, and aggressive salesmanship. IBM managed to prosper even through the lean days of the Depression, due to its equipment rental program, and was expertly positioned to reap the benefit of President Franklin Roosevelt is signing into law the Social Security Administration Act of 1935. (IBM revenues went from around $18 million in 1930 to $38 million in 1935. In contrast, NCR's revenues dipped from $58 million in 1929 to $16 million in 1932.[7]). By this act, Roosevelt altered not only the sense of security of a hard-pressed nation, but secured the near-term future of the fledgling data processing equipment business, for the mammoth information processing requirements in supporting such a system.

If expansion of the nation and of government were growth opportunities for the fledging data processing industry, World War II brought expansion and finesse to the technology. Department of Defense–sponsored research and development resulted directly in the post-war commercialization of several critical technologies that set the stage of overall advancement of the industry for decades to come. Core magnetic memory, the UNIX operating system, ARPANET—which evolved into the Internet—and the TCP/IP communications protocol were all by-products of the war effort. Dr. J. Presper Eckert and John W. Mauchly's pioneering work in the development the Electronic Numerical Integrator Computer—or ENIAC—for advanced calculation of ballistic equations was the watershed event that launched the electronic digital computer era. ENIAC, used in development of the hydrogen bomb, provided a quantum leap in computational horsepower, proving a thousand times faster than the previous electromechanical device.[8]

After the war, Eckert and Mauchly went on to develop UNIVAC, the first commercial electronic digital computer. Weighing eight tons, with over 5,000 vacuum tubes, UNIVAC was the first of the truly "big iron" machines to be harnessed for peacetime purposes. Remington/Rand, who acquired the brilliant, but highly undercapitalized Eckert and Mauchly firm, moved four-square to the head of the class in the rapidly evolving post-war data processing field. Remington/Rand's shipment of the first machine to the U.S. Census offices in Philadelphia in the spring of 1951 effectively "scared the hell" out of Thomas Watson. The deal had caught IBM flatfooted in one of its own key markets, but more importantly, UNIVAC's advanced tape drives threatened the lucrative 80-column "IBM card" business at its core.[9] (Remington/Rand, no doubt, took special pleasure in this, for IBM's lock on the 80-column standard, introduced in 1928, had foiled Remington's 90-column card standard, which was more difficult for clerks to handle and read.[10] For its day, IBM's 80-column card standard was as dominant a factor in the market as Microsoft's brilliant DOS triumph.)

This potential threat—as much as any other factor at the time—lit a fire under IBM. IBM followed with a reorganization of the company for improved worldwide reach and vastly stepped up its R&D efforts, creating the engine that would power the company's pervasive push into every market segment and niche in every corner of the globe for the next thirty years.

A spate of new machines was developed and released through the 1950s, setting, resetting, and constantly raising the bar for all contenders. The introduction of the 650 Magnetic Drum Calculator, the first mass-produced computer, in 1954 converted 10,000 IBM manual card processing sites into computer prospects virtually overnight. (The use of a four-inch diameter magnetic storage drum, running at 12,500 rpms, permitted a 2.4 milliseconds read/write access time.[11]) The 650 was followed significantly by the RAMAC 305 (Random Access Memory Accounting Machine), the first totally dedicated disk-based system. The 305 was hugely popular in manufacturing, for it vastly streamlined the manual sort effort of inventory cards every time there was a status change. Multiple cards had to be maintained on each part, each one pulled, updated, and refiled in large drawers—or "tub files"—that filled acres of counter space. With the RAMAC 305, each rotating magnetic disk in the storage unit held 100 tracks per side, each track capable of holding 500 alphanumeric characters. For its day, this was a huge advance—though today this could be held on a single chip![12]

IBM surpassed Remington/Rand in total number of units installed in 1955 (Remington/Rand commercialized UNIVAC) and had introduced its new workhorse, the 1401, but the market was under fierce attack with new machines introduced by Burroughs, UNIVAC, NCR, CDC, and Honeywell—known collectively as the BUNCH, with the emergence of Digital Computer Corporation and its DEC PDP-1—inaugurating the age of the minicomputer.

I, for my part, saw myself—as did my employer IBM—as a front-line troop to build and maintain our position. It was as auspicious a point to be engaged in American business as any since the founding of the Republic.

2 Assumptions—Truth and Consequences

One gallon of gas used in an engine of ordinary efficiency will do equivalent work of about 90 men or 9 smallish horses for one hour.

J.G. Landels, *Engineering in the Ancient World*

American manufacturing prowess following World War II was based almost uniformly on mass production. The pent-up appetite for goods, coupled with the natural economies of scale inherent in so huge a single, united market justified reliance on make-to-stock (MTS) production as the most cost-efficient, effective form of manufacturing. Roughly 90 percent of all industry was organized in this manner, the exception almost singularly being the manufacture of large industrial products with extensive lead times.

Manufacturing as a business—manufacturing with a capital "M"—had become Big Business. Operations were closely monitored through the short lens of prudent financial management. Manufacturing with a small "m," the activity performed on the shop floor, was a cost to be squeezed. Labor and materials were expendable. The relationships between management and labor, and management and suppliers were almost universally cast by management as adversarial. Workers were laid off and vendors played one against another as deemed necessary in pursuit of cost efficiency.

I found manufacturing—that which took place inside the four walls of the plant—a dynamic arena of fascinating intellectual rigor. I'd worked in production in a one-man print shop during high school and in a machine shop during leaves from West Point. Before joining IBM, I had worked at Bowl-Mor, where I had gained more experience in inventory control and production scheduling. The idea of getting your hands dirty "making stuff" that somebody *needed* was immensely appealing.

The high unit cost of processing data by computer, however, virtually required that IBM and other computer companies focus on promoting the general utility of automating simple tasks that were performed on a massive scale. In manufacturing, order processing and inventory management were both composed of components that especially fit this bill. More significantly, they were also both financial system requirements and of keen interest to corporate financial officers. The legions of clerical workers required to maintain control in these two departments made them obvious centers for automation. My job as a manufacturing system engineer was to assist in validating the application of computers to these areas, assessing, in part, how much computer power was likely required to offload repetitive manual tasks in those departments. After the client had signed the contract, my focus shifted to helping the manufacturer properly structure the data and programs to deliver on all the claims we had made to get the order.

In addition to the good fortune of being assigned to the Worcester branch office, I was relegated to working almost exclusively with manufacturers of industrial products. Then—as now—the market for large, complex industrial products was driven predominantly by real customer orders, rather than a forecast.

Most of the companies I worked with had rich legacies owing back to the advent of the Industrial Age. Crompton & Knowles built (and still builds, among other things) textile looms based on patented designs that gave birth to the American textile industry. Norton & Company manufacturers grinding wheels based on the innovation of F. B. Norton, an early Worcester settler whose all-emery composite grinding wheels, formed on a potter's wheel, revolutionized the craft, completely replacing emery belts, which were laboriously glued to the edge of wooden wheels. Norton grinding wheels were later augmented by grinding machines, earning Norton Machine worldwide renown and securing it a place in history for its indispensable contribution to precision in the manufacture of such vital elements in railroad rolling stock, and later the automobile, as axles, crankshafts, and gears.

The products manufactured by the companies I worked with could be classically characterized as make-to-order/engineer-to-order items (though this nomenclature came much later). Order and material management were critical functions. Capacity and labor could usually be augmented by subcontracting and overtime; but material availability—or more pointedly, material shortage—always threatened to bring production to a grinding halt.

Order and inventory management were intractable problems. The plant floor was continually whipsawed between changing order priorities, resulting in mammoth build up of work-in-process, prolonged queues, and misplaced, missing, and "borrowed" parts. Squadrons of expeditors worked to "hot," "hotter," and "hottest" dispatch sheets, aggravating both costs and quality. Late orders and unhappy customers were par for the course. When the power of computing was first considered as a solution, management was looking not so much to do new things differently, as to do old things faster.

The reorder point (ROP) method was the prevailing practice for ensuring adequate inventory to keep the plant running. Devised by Bell Laboratories in 1915, highly flawed though it was, the reorder point method provided a workable solution for managing parts inventory to protect against costly stock-outs that could shut down the line. The goal was to find some semblance of balance between too little and too much on-hand inventory.

ROP is basically keyed to a minimum stock-keeping level, or threshold, determined by a combination of historic part demand against standard delivery lead time and policy, augmented by an additional factor considered "just-in-case" safety stock. The higher the value put on customer service, typically, the higher the safety stock quotient.

Coincidental to the ROP threshold was the product of a calculation for determining the most economic lot size, or economic order quantity (EOQ). The EOQ was basically a square-root equation designed to balance the cost of activating a purchase order with the cost of carrying on-hand inventory, providing the best overall economic return to the company. In that both of these premises prove logically flawed under careful analysis, the resulting EOQ equation is a square root of two false assumptions.

Within the EOQ were such factors as setup costs, unit costs, carrying costs, and annual usage. In order to work as designed, the EOQ required acceptance of a series of assumptions. The first was that the future would repeat the past in fairly uniform fashion, with a reliable and steady depletion of stock on hand. EOQ was limited further in that it should only address independent, or end-item, demand as opposed to component or lower level parts. An assumption more deeply rooted in EOQ was company policy for how to value cash reserves; this had direct impact on affixing the cost of carrying on-hand inventory. The equation was further skewed by the fact that all manufacturing "fixed" overhead costs were allocated to production orders as burden, which tended to drive up the EOQ in order to keep setup costs down. This consequently tended to result in the build up of work-in-process, as burden could be spread further—made more economical—if lot and runs sizes were larger.

Flawed?

Most assuredly.

But given the sheer size of the task of tracking, tabulating, and calculating all the data points required to achieve greater granularity to the numbers, EOQ/ROP was considered more "scientific" than any other method then available for managing inventory. The convoluted set of assumptions propping up the methodology was hinted at, but generally swept under the rug as a necessity of the science, if the science was to work at all. Questioning basic assumptions was not top of mind in applying the new technology. As I said, we were primarily interested in doing old things faster.

I innately grasped the significance of applying computer technology to this whole complicated task, becoming a fervent convert to "in memory" processing over the repetitive tedium of manual calculation. Many others who were more seasoned, who had cut their spurs with IBM's more traditional line of unit record equipment, struggled with the transition. I was completely enamored with the potential for what this "marvelous electronic brain" could accommodate.

After nearly a decade of missionary sales efforts beginning with delivery of the first commercial computers in the early 1950s, the market was primed for something like the IBM 1401 processor. The transistor, invented in 1948 at Bell Labs, did not completely replace the vacuum tube at the heart of the computer processor much before 1960, which was when the IBM 1401 came to market. The 1401 was the first commercial, all "solid state" computer. Solid state engineering greatly reduced the footprint of computers in computing rooms and significantly reduced the purchase price of computer equipment. Not only was it powerful, featuring standard twin disk memory units that provided ten megabytes of memory, but it also had verve: the disk memory units could be easily mistaken for Wurlitzer jukeboxes, racking and selecting disk platters on command, setting and spinning the platter of choice to service the application of the moment. Like the jukeboxes of the day, they were mesmerizing to watch.

By the end of 1961, over two thousand 1401s had been sold. By the end of 1962, computer storage system sales at IBM passed accounting machine revenues for the first time. The 1401 could be used with punch cards or magnetic tape, run stand alone or work in tandem with other systems. The rental program started at $2500 a month. The popularity of the 1401 effectively cleared the cluttered market of several less successful market contenders.[13]

With core memory and solid state engineering, the 1401-class computers marked the second generation in commercial computing. Popular though they were, they still required an army of skilled application specialists to help customers custom program them for use.

In 1961, the Norton Company purchased a model 1410 (an enhanced 1401) configured with 50 megabytes of hard drive, which was ungodly huge for the day. The grinding machine side of the business used it to maintain inventory balances, with just barely enough capacity to handle orders. It kept the main files on hard disk, and the rest on tape drives. The grinding wheel side of the business—the largest producer of wheels in the world—could only use it, however, for order management, as the description of individual wheels, given varying diameters, edge-angles, composites, coarseness, and such, took more than 80 characters (the capacity of an IBM punch card) to fully capture individual product uniqueness. The hardware cost alone of Norton's 1410 was $3 million.

When the 1401-class machine first came out, it was devoid of a programming language. The task of giving it the intelligence to know what to do with all the data it was fed had to be laboriously entered via arcane machine terminology. Toward the end of its life, SPC and Autocoder, two rudimentary programming languages, were developed, modestly enhancing the machine's productivity, but it was still arduous going. And yet, it was generally conceded that the 1401 was a slick machine.

The bill of material (BOM) processor program, inaugurated on the IBM 305-class computer in the late 1950s, was heavily promoted as an ideal application for the 1401. IBM essentially "gave away" both the program and services to program the application in pursuit of moving Big Iron. "Free" or "bundled" software and services though they were, producing a bill of material processor program was no small undertaking. A basic, straightforward processing loop driven by a series of tables, the task of programming the 1401 for BOM explosion took a minimum of 12 months, more typically 18 months. Completion of this task was always cause for a celebration.

The BOM processor application was essentially a series of data tables that were "chained" one to the other by pointers to other records. The table that held the descriptions of all parts (the parts master) was chained to the table that described the next level of parts and components that came together to make a parent in the product structure. Each component part structure in the entire bill was represented by a single punch card. This card contained, at a minimum, the unique part number that identified the part, the parent assembly number (of the assembly where the part was used), and the quantity required. These cards were used by the BOM program to create the product structure file.

The whole structure (hence "bill"), from a single finished end item down through all the levels of component parts, each level branching ever deeper into lower level parts and assemblies, had to be created with absolute fidelity in the program for the application to work. The entire program filled dozens of large boxes with punch cards. But getting these all properly punched, or "coded," merely got you half way home.

These cards then had to be loaded into the computer's database structure so that when you wanted to "explode" a bill based on an order to build a certain product in a certain quantity, the BOM processor could properly calculate what material you needed. As you might imagine, this explosion of the bills (each part at each level its own shop order) produced reams of paper reports so voluminous that they had to be delivered to material managers on wheeled carts.

At Norton in Worcester, the closest 1410 was located at regional headquarters in the Time-Life Center in New York City. When we were ready to test our programs and were loaded down with our many boxes of cards, we scheduled our journey to Manhattan. We traveled on a Monday to be sure we were present in the city, ready to test by the end-of-day Tuesday. From midnight Tuesday, we had time, at most, for three processing runs. If we were blessed, we loaded the cards and ran the program to completion on the first pass. More typically, we stood by until the procedure was seized by an error or a miscued card. One wrong card would halt the validation of the entire run. When that happened, we had to investigate and resolve the error. After we had successfully run all cards through the reader, we performed a memory core dump, committing all positions to voluminous paper-based printout. Come 6 A.M.—whether we were through or not—we had to pack up and get out of the building. By day, the domain belonged exclusively to the crunching of financial numbers.

The value of serendipity in life is often not revealed immediately. Working with a select list of clients in Worcester in the 1960s, all, more or less, characteristic of make-to-order/engineer-to-order production, I was something of an odd duck at the time for the expertise I was garnering. The vast majority of American manufacturers were make-to-stock (MTS) operations, driven by forecasts. MTO/ETO is powered on customer orders. The difference between the two modes of production is significant, causing a profound ripple effect that touches almost every aspect of managing the shop, from purchasing, lot sizing, scheduling, sequencing, cycle time, throughput, and turns, to invoicing, cash flow, and ledger management. Never having been one drawn to conformity, following a different drumbeat in industry was appealing to me. Being a big fish in a small pond definitely has its career advantages—unless the pond dried up. Time and history would prove on this count, however, to work in my favor.

Faint, hairline fractures were already beginning to appear in the MTS paradigm, particularly in the automotive and electronics industries, though not to any degree yet to sound wild alarms. That would come later in the 1970s and 1980s, after foreign competitors secured their beachheads at the low end and could begin their inexorable march toward the high ground. This was particularly true of the Japanese, Germans, and Italians, who were all still rebuilding their infrastructures during the 1960s. The advantage of losing the war (an odd, though certainly not spurious notion), with the subsequent destruction of their industrial base, was that "foreign" competitors had a fertile field for experimentation and innovation in manufacturing methods. To their advantage, their markets were far less geopolitically oriented to mass production as it was practiced in the United States. With traditionally smaller local markets, smaller production batches aligned more toward a configure- or make-to-order approach made more sense. Agility and nimbleness were critical. For Americans, our weakest link in the ramparts of our production style was in the area of quality. It was an area where these competitors-in-waiting would focus with laser intensity, widening the crack over time with the wedge iron of consistently better quality.

Questioning the assumptions behind the strategy is always a difficult feat to achieve in its entirety. Being intelligent beings, we are fairly adept at prying into the first level of basic assumptions that support the general structure and order of things, and even at exposing the second, underlying circle of assumptions. But it is extremely challenging to pass beyond this layer, penetrate the belt of protective assumptions, the "negative heuristic,"[14] that commonly is accepted as sacrosanct and inviolable, and prevent revolutionary breakthroughs.

The narrow field of American industry organized around the MTO/ETO model worked to a significantly different set of assumptions. Though I had little appreciation for the full measure of the significance at the time, serendipity had thrown me, like the Uncle Remus tale of Br'er Rabbit, into the thick of the Briar Patch. Truth was, it was exactly where Br'er Rabbit wanted to be. In time, I would come to immensely value the good fortune that had flung me likewise, right where I needed to be.

During those first few years at IBM, I wrote several symposium papers that were sufficiently inspired to receive some attention. I also developed a 3-dimensional axis tool simulator for more accurately graphing viable economic order quantity sizes, adding the desired level of customer service as a deterministic percentage in the calculation that had historically looked at only volume of orders and capital costs. For this I received a Corporate Outstanding Contribution Award from IBM.

These activities served to elevate my visibility in the company, consequently earning me an invitation to corporate headquarters in White Plains in the spring of 1963. I was one of a select group of manufacturing industry specialists from around the country called to spend two weeks sequestered as a part of an auspicious conclave.

Nervous though I was at the prospect that I might make a complete fool of myself—only 29 years of age, joining more experienced associates—I was pleased, if not outright impressed with myself. While youth can be a liability, I was determined to make it an asset.

I took my full measure, kissed my family good-bye, and started down the road to White Plains.

3 Brainstorming in White Plains

The IBM System/360 provided roughly 1/1000 the power of a current generation 200 MHz laptop computer.

Dick Lilly

Only in Genesis does Creation move from the Void to Eden in less than a week. Software development starts from the same place, but takes longer. It involves many iterative steps; is typically done by committee; and, more often than not, it requires working on Sundays to meet the schedule.

But—keep in mind—it is creative. And usually inspired by lofty aspirations. Unlike in Genesis, however, the Devil is not in the Garden. It is in the details.

By 1963, IBM was confident in its belief in the potential of applying computers to business problems, though its future dominance of the market was anything but assured. The horse race among IBM and what were to become known as the BUNCH (Burroughs, UNIVAC, NCR, et. al.) for market position was still in the first furlong out of the gate, all parties jockeying aggressively for the inside rail. IBM was pouring massive resources into the effort, building out manufacturing capacity, putting more feet on the street, and beefing up brainpower in its labs. My trip to White Plains in the spring of 1963 was part of a critical plan to take the race well through the first turn and put IBM clearly out in front along the back rail.

There were about a dozen of us who had been selected from IBM's manufacturing services organization. We were quartered in a local White Plains hotel and arrived punctually each morning at 8:30, mingling amidst the three or four thousand other IBM employees flooding the corporate headquarters' parking lots, eager to get to work. There was little to distinguish us from the general crowd, all dressed alike in similar blue suits and white shirts. Our small group was ensconced in a nondescript conference room. But within that conference room, we were unique in our focus. Discussion was extensive and reasoned, but also impassioned, at times, colored distinctly by the slang and shorthand argot of manufacturing production and inventory control. We regaled one another with stories and challenged each another's precepts. Our charter was to complete a comprehensive design in two weeks, time that could serve as a master blueprint to an integrated manufacturing production information control system.

It was wonderfully exhilarating, especially for me, being the youngest member of the team. Though we were bound by our collective experience, we were encouraged to take full rein, for computing was so new nobody knew for certain where the outer limits truly lay. Our goal was to push to the limit without escaping the laws of reality and the then-state-of-the-art computational horsepower.

We were, like similar teams of IBM specialists assembled to consider the requirements of other key industries—banking, insurance, medical, and distribution—privy to information that teased as much as motivated us to push as far as we collectively dared. While we worked on chalkboards in our respective conference rooms, outlining the information flows that defined our respective industrial domains, computer design engineers were busy elsewhere scoping the details of a new computer. Not merely *a* new computer, but a *whole family* of computers. Our job was to prototype a comprehensive span of application functionality to match the expanded horizon being engineered in development of the IBM System/360 computer.

IBM was banking on its assessment that the market was ready for a new, general-purpose computer. Commercialization of the technology had crossed through the critical first dozen years. Computers had proven their feasibility, if not entirely in hard numbers in reduced clerical headcount, then certainly in the hearts and minds of the financial chiefs for their ability to rapidly crunch numbers and tally ledgers. We were a long way yet from commodity selling, but a boundary had been fixed delineating computers as a legitimate business expense for the Fortune 500. The premise of benefit had been established: computer technology aided greatly in the conversion of chaotic data into meaningful order. Order meant greater control. If not yet a religion, there was sufficient faith to start laying the foundation for the church. IBM's strategy was to use the System/360 to raise a fervor and fly Big Blue's flag from the tallest spire.

The System/360 was to be based entirely on solid-state, integrated circuitry. Bell Labs had inaugurated solid-state engineering with development of the transistor in 1948. Replacing vacuum tubes as the key switching devices for changing binary values promised a quantum leap in computer speed. Texas Instruments had fabricated the integrated circuit in 1959, though Robert Noyce of Fairchild Semiconductor received the patent, effectively commercializing silicon-based integrated circuitry some months later. The integrated circuit essentially permitted numerous transistor switches to be placed within one individual silicon substrate, effectively accelerating computer speed another quantum leap forward.

Additionally, IBM wanted to leverage the R&D it had performed in the 1950s on government-funded development of what was known as the STRETCH computer, originally built for the Los Alamos Scientific Laboratory of the U.S. Atomic Energy Commission. The STRETCH computer measured 6x30 feet. Its chief accomplishment was that processing was "managed" by a level of stored instructions, a set that comprised a viable operating system. Further, STRETCH employed an interrupt mechanism, a timer, and a supervisory mode, all of which permitted improved management of processing runs. It had standard input/output (I/O) channels.

These features would be augmented in the System/360 by advances in electronic memory, first pioneered by MIT in the 1940s in its WHIRLWIND computer, later commercialized by IBM and others in the 1950s in such machines as the RAMAC 305 and the 1401, making large-scale, reliable memory affordable.

Beyond the technological advances envisioned for the System/360, however, at heart it was conceived not simply as an assortment of processors and peripherals, but a collective of interchangeable, seamlessly compatible hardware components. In essence, it was a family of computers, ranging from a low-end unit, with from eight to 64 kilobytes of memory upwards to 1024 kilobytes—signifying better than a hundred-fold increase across the range.

This development, this ratcheting upward of the competitive bar, was what lay at the core of our purpose in gathering in White Plains. For our part, our mission was to raise the conceptual bar for software applications in order to boost sales of this new generation of hardware when it was delivered to the market.

What we were attempting had never been done before. No one had ever attempted to create a commercially available architecture for an integrated software system. Given the anticipated state-of-the-art of the technology, we focused primarily on what we knew we could successfully tackle and on what manufacturers were, in that day and time, most vexed by: materials management.

The problem was twofold: not simply how to plan, but *re-plan* and how to respond to the real-case scenario of manufacturing, characterized by the relentless occurrence of unplanned, unanticipated events. In a perfect world, plans work perfectly. But in manufacturing, events begin impacting the plan almost from the moment it is prescribed. Management changes its mind about what the plant is supposed to be working on. Engineering improves product design, altering the bill of material. As the future draws ever closer to the present, sales is forced to tweak, change, and/or scrap the forecast for another set of projections. These and a thousand-and-one other changes, many done with the best of intentions, wreak havoc on the life expectancy of any one plan.

One of the historic, major drawbacks to the reorder point method of inventory management is that it is based solely on data that is ... well, *historic*. It also looks at parts inventory as ... well, *parts*, not whole products. The promise of computers that fired our imaginations was that we might be able to move to a more *future*-oriented model, one that considered *whole* products. The closer we could approximate these dynamic parameters, the better manufacturing management could efficiently and cost-effectively avoid problems created by material shortages. Stock-outs during a product run are extremely costly, resulting in inefficient setup changes, escalating expediting costs to meet order deadlines, and spiraling work-in-process buildup.

We set out to specify the logical flow of information supporting the manufacture of finished goods in a make-to-stock environment, this being the predominant mode of production at the time. Working at our square conference table, mapping the flow on the chalk board as we went, we started with sales forecasting and successively moved through engineering data control, inventory control, requirements planning, purchasing, capacity planning, operation scheduling, and shop floor control. We argued over definition of terms, requisite policies, mandatory procedures, and chained sequences. Little escaped our proclivity for debate. We would belabor a point to the point of the ridiculous, sometimes achieving breakthrough, sometimes merely growing weary, deciding to table the issue and move on.

Reticent at first, I warmed to the challenge. I, like everyone else in the room, believed we were participating in something pivotal and momentous in marketing to American manufacturing. Hardly solemn in decorum, we nonetheless took our task quite seriously.

Some functional requirements we knew to be beyond our grasp. Where we encountered gaps and chasms, we tried to post warning signs for those who would follow.

By necessity as well as the charter IBM had put before us, we had to establish and accept certain assumptions in order to make any progress at all. The ruling assumption was that promoting the sale of IBM hardware was a premise without equal. On the face of it, it seemed an obvious and perhaps innocuous point. Additionally, we had certain hardware features that served as design constraints in what we had to do, primarily the speed and memory of the new System/360 family of computers. Revolutionary as the System/360 was, embracing a wide range of design elements fostered over the years, it still had constraints that circumscribed solutions to production problems we had to address. Time would reveal that our cleverness of words and semantics in accommodating the assumptions we made had a dark side, one that would haunt the industry in terms of stifling real intellectual rigor and advancement of the discipline for twenty-five years to come.

Chief among the known gaps we encountered were three sizeable crevasses. We sought to address each of these by creating work-arounds with impressive sounding names that could be easily marketed: *level-by-level bills of material, standard leadtimes*, and *infinite scheduling*. The three are integrally linked, with the first work-around (level-by-level) requiring that we engineer the other two to support it; and the last (infinite scheduling) having the effect of greatly compounding the negative impact of the whole notion of standard leadtimes. The full devastating impact of what we were unleashing, however, was to dawn on me only years later.

Given the limitations of computer disk storage, memory, and available tools, it was impossible to entertain the idea of creating a program to store a company's process sheets, which included the multilevel bills of material, their operations, and their specifications. The work-around we devised began with isolating the bill and ignoring the operations and specifications. The multiple levels of the bill were further isolated, one from the next, creating a *level-by-level bill of material structure*, where every part, subassembly, assembly, and parent became a separate and individual bill. This had the advantage of creating no adverse impact on calculating costs at each level, which was critical in terms of winning approval of corporate finance officers during the sales cycle. But the resulting impact on management of production in the shop was mammoth, in that the number of shop orders required to build a product grew exponentially. Furthermore, there was no simple continuity or connecting thread maintained among the various shop orders required to manufacture a single end item for a specific delivery date.

Standard leadtime was a mechanism we devised to address the coordination of bringing the various levels together in some semblance of sequential order. Each of these level-by-level shop orders had to have a designated start and finish date assigned in order to be "managed" in the shop. These dates were generated based on a variable we defined as the standard leadtime in the Parts Master File, typically specified in days or weeks. As anyone working in manufacturing well knows, however, "standard leadtime" is an oxymoron. Leadtimes for a part vary from order to order based on material availability, current priorities, the existing load in the shop, and the occurrence and severity of bottlenecks. For the construction of the level-by-level bill of material structure to work, all this had to be ignored. This was easy enough to accommodate, as these factors were unknown and unknowable given the state of technology of the time. The value inserted in the standard leadtime field was routinely selected by someone in data processing. In time, the cachet that became associated with "standard leadtime" would be polished to a bright luster under the skillful craft of the IBM sales team and accepted as sacrosanct by the corporate financial officers and the data processing department, such that far be it from anyone in production to question the validity (some might argue sanity) of its use.

Infinite scheduling was created to address problems inherent in level-by-level and standard leadtimes. Foremen on the shop floor were now inundated with paper reports for shop orders generated by the level-by-level bill of material explosion, with standard leadtime values for completion dates that had no basis in reality. To provide some relief to these problems, we came up with "infinite scheduling" as a means of providing production management with a means to establish priority for operations in the shop. Infinite scheduling, in reality, was neither scheduling nor infinite, but only a means of dating operations by a process of backloading the shop to establish suggested operation start dates that the shop could work with. The backward scheduling calculation (starting with the due date of the shop order and working back through each operation to determine the start dates of each) was "infinite" in nature in that it was done not only without regard to the existing load of work already in process in the shop, but also, and more importantly, without regard for prior shop orders already assigned to the same resources and dates. Further, it failed to reckon with any information regarding resource bottlenecks.

At the time, we understood some of the dynamic problems intimated in both standard leadtimes and infinite scheduling, for we knew that the far superior solution lay in finite scheduling. But we knew that finite scheduling was completely beyond the capability of the System/360, even on a completely dedicated machine, thus we knew that the challenge was to devise some other means to accommodate the idea of finite scheduling, if in fact we could not readily resolve it. As we came to the end of the next to last day, the group leader encouraged us to be creative in our dreams to see if a solution might be found. By now, everyone's creativity and patience were severely tried and we were eager to go home!

I was the one who put forth the recommendation we would follow the next morning. There was, in fact, nothing substantive we could do, I advised the group. The 1401—even the System/360—simply wasn't fast enough. But I had an ingenious—I thought at the time—way to get around the problem. It was something that I'd picked up from Robert Goodall Brown, an Arthur D. Little consultant. Brown had espoused the idea of *exponential smoothing* at an IBM training seminar in Cambridge, Massachusetts, in a presentation on IMPACT, a novel approach to the management of material in distribution. I offered it here. The concept was little more than a fancy way of handling weighted moving average.

The idea as applied to scheduling involved smoothing the queue at every work center. In brief: an average queue time is calculated over a set length of time, and period by period, adjusted, based on a rolling average, against a percentage offset calculated from the period just completed. If this is graphed over time, what we get is an exponential curve, hence, exponential smoothing. It serves to get you in to the ballpark—if you can fathom your way through the mental gymnastics and slippery semantics. Slick as it seemed, what time would show was that it was too meaningless for the average production planner to fool with. But it got us to the close of day Friday with a neatly finished blueprint.

If we'd been bound by the rigors of working inside a mathematical theorem-even a mathematical conjecture, for that matter—we would have been susceptible to attack the moment we walked out the door. Our objective, however, was nowhere near as demanding. We were chartered with merely driving a stake in the ground. In truth, it was not meant to serve so much as a system schematic as a blueprint for successfully marketing the System/360. A blueprint, we had, and having that, we all were more than happy to pack up and get out of town.

4 Shift in Values

Experience has shown that manufacturing has a need for

1. a central information system, and

2. a framework to facilitate mechanization.

Production Information and Control System, IBM PICS manual (Courtesy of IBM Archives.)

Through the end of 1963, the computer hardware playing field was not tilted in IBM's favor. The company's modest lead in the market, based on the success of the 1401 processor, was anything but secure. New competitors were entering the market and numerous new machines were being rolled out from old and new competitors alike. In December, Honeywell announced its low-cost H-200 machine. Marketing of the new machine was augmented by marketing of a companion program, code-named "Liberator," aimed directly at IBM's installed 1401 customer base. Liberator was built around a special translation program for easing customers away from the more expensive and slower 1401 processor. Moving through the first quarter of 1964, IBM reported 196 losses to the Honeywell initiative.[15] Honeywell and the rest of the emerging BUNCH saw a promising future, indeed.

That was before April 7, 1964, and the announcement of the System/360 computer. The announcement stressed that the System/360 was not simply "a bunch of processors and peripherals, but an aggregation of interchangeable hardware units with program compatibility top to bottom."[16] It included the unprecedented delivery of six different processor models, from the Model 30 at the low end, with 8 to 64 kilobytes of memory, ranging up to the Model 70, with as much as 1024 kilobytes. Rental rates ranged from $2,700 to $11,500 a month. Additionally, the announcement included 44 new peripheral devices.

In the first month following the announcement, IBM wrote over 1000 orders for the new machine.[17]

The *Production Information and Control System* manual was a companion piece in the IBM marketing arsenal for the System/360. It was only a manual, a rather terse 100-page document, equal parts vision and prescription. Nobody—anywhere, in any form or manner—had implemented such an integrated system. A few companies had implemented pieces of the inventory and MRP "modules" on their own, but not in the specific prescribed manner outlined in the PICS manual. To call the document a system design specification as we know the term today would be charitable. Its purpose was to outline, in broad strokes, a vision, to paint a picture of the possibilities of a systematic approach to the automation of production information.

From the preface:

> This publication enables the reader to visualize the management of a company as a total system. But in addition, it provides new knowledge of subjects seldom discussed before
>
> - value of common data files
> - flow and interaction of manufacturing applications
> - workings and use of transaction entries
> - techniques for disk file organization
> - use of symbolic labels to define DATABASE records

Elsewhere in the preface, we read:

> This manual, by defining the applications that make up a production information and control system, paves the way for a manufacturer to convert to mechanized production control.... Several factors permit solutions to the problem of mechanization in this area: 1) the IBM bill of material processor program, which organizes disk files and maintains the record data; 2) the enhanced speed, flexibility, and capacity of IBM's direct access storage device on its System/360 computer; and 3) the IBM

operating system program concepts with the ability to maintain continuity between jobs.

The production information and control system is a logical and orderly growth plan for a manufacturing organization to do a better job of managing men, machines, material, and money. The goals are clear:

- Increased productivity
- Increased profitability
- Improved management

The system can grow as the user grows. And the user will obtain tangible results long before the total system is installed. (Courtesy of IBM Archives.)

Good thing, too, in that there was no "total system" to install. But no need to quibble. The System/360 was hot. Production in the IBM Poughkeepsie plant could barely keep up with orders. The System/360 became the first killer machine of the information age, with orders flowing uninterrupted like a river at flood well through the 1960s.

And they weren't cheap, either. The Model 30 (a thousand-fold less powerful than the average laptop today) started at close to a million dollars—equivalent to around $16 million in 1999 dollars. A 7.5 megabyte hard drive cost $22,000; 16 kilobytes of memory cost $56,000. (To give you some perspective, my salary in 1963 was only $10,000.) Only the Fortune 2000 could afford them. But buy them they did—in volume.

One of the key selling points of the System/360 was that, for the first time, an extended architecture created investment protection for the vast expense companies had to pour into software development in order to run them. Theoretically, you could start with a Model 30 and migrate through the whole range to the Model 70. This made the idea of considering a software investment of the size required to develop a full-blown PICS integrated application as something other than total insanity.

Hot as it was, the success of System/360 was not without costs. The runaway success of the new machine was responsible in large part for piquing the interest of the U.S. Justice Department to delve again into the business practices of IBM. An investigation was launched in 1966 that would consume thousands of man-hours, diverting company resources and attention, and siphoning off millions in lawyers' fees.

The machine's explosive proliferation also created a huge demand for application software to run on it, a commodity that was all but nonexistent when the machine first came to market. Small independent development firms sprang up, oftentimes sprouting as entrepreneurial spin-offs from corporate project teams involved in developing in-house, homegrown systems for the Fortune 2000. While the lawyers wrangled, salesmen booked orders, and programmers worked furiously to stay acrest of the tsunami of demand for applications.

The blueprint outlined in the PICS manual served as the basic roadmap for traversing the manufacturing software terrain for the next forty years. True, it pointed the way over little more than a single-track road in many places, and completely ignored whole corners of the landscape. But again and again, you only have to thumb through old system documentation, industry journals, conference proceedings, trade press, and marketing materials to plainly see the ghost image of PICS replicated repeatedly right down to the cusp of the 1990s.

Here again, what was gained was not without cost. Though few companies would implement much more than financials and inventory for the next fifteen years or so, the impact of just these two functional "modules" was sufficient to stress a whole generation of material managers. Gone were the comprehensive tub files filled with cards that provided complete material pegging to orders. Yes, you could, for the first time ever, do a "where used" query against a bill of material to assess the impact of an engineering change order, for example. But no, you could not do a historical look-up to see where, in fact, specific parts had been consumed in the building of real orders. You couldn't determine whether items had been received, reported as defective, or scrapped. Nor could you determine whether they had been issued to an order or whether any adjustments had been made.

The idea of PICS, however, did appeal immensely to the people in the finance department, the people with the authority to write the check for it.

During the sales cycle, overall control was almost always granted to the financial accounting arm of the business. Accounting's inquiries regarding cost of material, labor, and burden, were always answered solidly in the affirmative by the IBM sales team. That typically proved good enough reason to buy. The boys down in the shop, after all, were not even allowed in the building that housed the computer. As time would prove, manufacturing management would have to wait well into the 1980s before "progress" touched their lives in any significant fashion.

There were trouble spots elsewhere. In one instance, the result was very near and dear to manufacturing at Big Blue.

A sales forecast typically looks into the future with any level of detail out through only a coming twelve-month window. In terms of system design, this gave twelve-month-sized buckets for depositing manufacturing and customer orders for finished products into available time slots. The manufacturing lead time on the System/360 was set conveniently at 12 months. Much of the requisition, acquisition, assembly, and test of component parts did, in fact, fall within a 12-month window. But there were numerous lower-level components that extended that window well beyond a year, out 15, 18, and even 24 months. Because there were only 12 monthly buckets to a forecast, when the bill of material calculation was processed, the requirement on these items came up as ... ***zero.*** As in "*no requirement.*" To make this respectable, I suppose, this went by the fancy name of "lead time tail-off."

Somewhere around 1966 or 1967, the Vice President of Manufacturing, whose Poughkeepsie plant was responsible for building the System/360, came to understand that he had a real inventory problem on his hands. No small item, the problem was in the neighborhood of three-quarters of a billion dollars in excess inventory. This was directly accrued against his annual profit and loss statement, and it seriously made him look rather slack in managing his operations.

His solution, he decided, was to call in a busload of Big Eight accountants to scour the books to pinpoint the problem. Frank Carey, then the savvy President of IBM's Data Process Division (later IBM Company President), stepped forward and suggested rather than the expense of hiring outside consultants, why not let a small group of IBM manufacturing system specialists take first crack to see if they could come up with a recommendation to rectify the problem.

I was one of the specialists invited to Poughkeepsie. It was enlightening. Faulty information system design, coupled with contradictory management edicts, can carry you a long way down the well-paved road of good intentions. And we know where that leads.

There in Poughkeepsie, adjacent to the sprawl of the production floor for assembling the System/360, the hottest selling computer of its day, was a room the size of a large school cafeteria. Like students devouring lunch, hunched over row upon row of IBM gray metal desks, sat a legion of inventory control specialists. Each individual had a select list of parts he or she was responsible for. They worked feverishly under the edict that the buck stopped on their desks if any of the parts on their lists were ever short, causing a production stoppage. Many of the parts on numerous lists fell into the black hole of "leadtime tail-off." Rather than schedule these requirements separately, however, each stockpiled in his desk a reserve of all his parts that might stop work on the line.

The proliferation of parts—of *extremely expensive* parts—was staggering. It was the same as if all these specialists had stuffed wads of cash in their drawers—out of sight, but purposefully near at hand. I happened to point out the fact that leadtime tail-off was directly responsible. The observation was passed along, up the chain. Nothing much was done about it. We finished our engagement and went back to resume our regular duties.

In the spring of 1967, I was invited to attend one of the national biannual IBM industry conferences. The leading names in manufacturing information systems were in attendance. Prominently present were Dr. Joseph Orlicky and Oliver Wight, veterans of years of experience on the shop floor, former IBM system users, now both employees of the company. These gatherings were typically well-attended, popular conclaves—part seminar, part pep rally. I had been to numerous others before. They served as a break from the routine of work in the field, as well as an opportunity to retap the well of creative ideas for inspiration to spur another six or twelve months of dedicated effort.

This session was different, at least for me. And significantly so. It began to weigh upon me, sitting in the audience over the three or four days I was there, listening to round after round of presentations, that either the industry had made a subtle shift or I had. One or the other, perhaps both. I began to grow aware of a sense of dismay in what I was hearing. I couldn't shake the feeling. The benefit of information technology in aiding manufacturing management was clearly touted; that it promised to make it more cost efficient, more competitive was not questioned. But the emphasis was distinctly elsewhere. It rested squarely on moving hardware. Not surprising perhaps, but what really disturbed me was the sense that we were not to let implementation of a solution get in the way of a hardware sale. The two might be related, but they were distinct. And it was clear which one got first billing. Always.

I'd never been confused about what business IBM was in. IBM—then, as always—sold hardware. They sold it aggressively and creatively. And there was no question that there was real value in the proposition for the customer. But this was a new twist, in my mind anyway. My whole focus was on helping customers realize the value proposition in the equation. To question the importance of that, to subjugate it so clearly to the moving of Big Iron, didn't sit well with me.

I left that meeting with much on my mind. I was hardly energized, at least not in the manner such conclaves are meant to inspire. I look back now and can see that that session stood as a watershed event. And not just in my career. But as time would prove soon enough, we were at that point on the edge of a new era. Software—independently developed and supported separate from the domain and dominance of hardware sales—was a reality, if not fully borne, at least well into the cycle of gestation. The idea of it was too powerful, even at this stage in the history of commercial computing, to be denied. In my mind, software was where the true value lay. Now—and in the future.

5 Independence and the Independent Software Industry

Commercial designers go where the perceived market leads them.

Stewart Brand, The Media Lab

I quit IBM on April 15, 1968. I had four children at home, one grievously ill, requiring a live-in nurse. Giving up the security that working for a Fortune 500 company offered in those days required more than a little faith. But I believed the timing was right and the decision was sound. I was smart enough to know, of course, there were no guarantees.

I left IBM to form a partnership with Bill Watson. Watson and I had become acquainted while both of us were assigned to IBM's Boston District offices, where Watson worked as an educational consultant and I was a manufacturing specialist. Watson was very professional, very polished, poised, and confident as an IBM salesman working out of the Concord, New Hampshire, area prior to being assigned to Boston.

Through the early months of 1968, we both had been giving thought—independent of one another—to starting in business on our own. The market was crying for talented, experienced people to work with companies implementing computerized business systems for the first time. There was not enough talent to fill the need at IBM, let alone elsewhere in the industry. American business was desperate to the point that anyone who espoused experience was almost instantly employable. The opportunity this presented was too great, prompting some of the big accounting firms to launch their first initiatives into computer consulting services. From the point of prudent business practice, this made absolutely no sense to me, that you would ask those who are chartered with auditing your business practices to become critically engaged in a hands-on fashion with the workings of your business. The accounting firms, however, were confident in their own minds that this represented no conflicts of interest and pushed ahead aggressively. Watson and I both viewed this as evidence of how dire the need for help had become. Sharing our thoughts on the issue—and our independent musings on starting in business on our own—we decided the right combination of talent and skill held a better-than-even chance of success.

By this time, I had extensive system engineering expertise, and Watson had the field sales experience and financial wherewithal necessary to give a serious run at starting a viable business. I had a contact at a manufacturing company who was willing to contract with me for one day a week for a year to have me help get its manufacturing system up and running. The work would serve as a foundation for us while Watson solicited additional business. That spring, we both resigned from IBM to start Manufacturing Management Sciences. We flipped a coin to see who was going to be president. I won the toss.

George Sontag, data processing manager for Heald Machine, a division of Cincinnati Milacron, signed us to our first contract. At the rate of $300 a day, guaranteeing one-day a week for a year, I agreed to design a manufacturing control system for Heald's Worcester facility. Though I had done numerous similar projects while at IBM, this was the first one that I was to design entirely on my own. Watson and I were thrilled to be in business for ourselves, and the challenge of working without the safety net of a benevolent corporation behind us only served to fire our will to succeed.

Our second job was with International Packing in Bristol, New Hampshire. At International Packing, Watson developed a system for scheduling ovens used in the manufacturing of packing seals. That was followed by a contract with Converse Rubber to develop a bill of material system to drive an inventory system, with a commitment for a general ledger system to follow. The rapidity of our success forced us to be versatile; Watson both sold and implemented the application at International Packing, and I both sold and engineered the work at Converse Rubber. At Converse Rubber, we were paid $17,500 for the BOM system and another $17,500 for the GL system, and we managed to retain the rights to the programs. In time, we would sell over 5,000 general ledger packages based on the original design we did for Converse Rubber.

Shortly after the anniversary of our first year in business, IBM, under continuing pressure from the Department of Justice, announced that it would no longer give away its software and services as part of a bundled deal with its hardware sales. Henceforth, IBM would "offer certain system engineering activities, most customer education courses and many future computer programs for a charge."[2] The "unbundling" announcement, coming on June 23, 1969, was made primarily as an outgrowth of formal complaints to the Justice Department by Control Data Corporation of Minneapolis, Applied Data Research of Princeton, New Jersey, and others against IBM. The June announcement by IBM legitimized the range of fees we were then charging, but more significantly, marked the definitive start of the independent software industry.

Other than the IBM PICS system—such as it was—and a program offered by Honeywell, there were no other packaged manufacturing systems in the market. The ICP Directory, started by Larry Welke in 1967 to track the nascent software market, listed 485 different programs available from 150 different companies across the full spectrum of business applications. The majority of these offerings were, in fact, user-developed, or in-house systems, primarily financial and payroll sytems.[18] Opportunity became the byword of the fledgling industry, and anybody with an inkling for independence threw a hat into the ring.

These were wild times. There were more than a few operations that were pretty sketchy in the details of their credentials. A lot of "smoke and mirrors" was sold as software. It didn't take much more than a business card and a sales pitch to set up business. The standard practice, it seemed, was to sell yourself as knowledgeable whether you were even vaguely familiar with what was being requested, quote a price and delivery date, then hope like hell you could pull it off before you ran out of money. If you were lucky—and not everybody was—you broke even. If you were really lucky, you broke even and retained the rights to the software. And if you were extremely lucky, the software was something that you could learn to remarket without a lot of reinvestment in development time.

Watson and I were extremely lucky. With our "package" from the Converse deal and the unbundling announcement, we were doing a land office business in sales of financial software systems. We decided to change the company name at this point to differentiate our custom engineering consulting business from our package sales. We reserved Manufacturing Management Sciences (MMS) to identify our consulting business, but began doing business in software licensing as Software International.

On the MMS side of the house, we developed a reputation for getting customers rapidly up and running on the IBM Bill of Material Processing program. This we would guarantee to do over the course of a weekend—an unheard of accomplishment. We did this by having our customers accept a certain level of "standardization" in the BOM program, such that we could reuse much of the batch card programming we'd deployed at our previous client site. We used a standard set of parameters in the cards, changing only those variables customers felt were absolutely necessary to accommodate their requirements. For this, we charged between $6,000 and $7,500—a fairly steep fee for that day. But we also guaranteed a "knowledge transfer" for maintaining the program; we did this by requiring that someone from data processing and at least one end user be a part of the weekend installation team, working side by side with us. Our customers easily recognized the value in our methodology and gladly accepted the price structure.

We also devised a method to take advantage of the disk storage cylinders in the System/360. By creating a COBOL chain file in our program, we were able to engineer variable-length fields (not previously available in the market) that provided greater flexibility in programming as well as speed in processing execution. This was especially useful, for example, with accounting programs. It enabled our programs to work from a single customer master file, but one that was chained to multiple individual invoices beneath it. This design feature, coupled with the programmable disk storage "read" technology in the System/360, made it possible to set the read head once and have it read through 20 different tracks without moving the head again, gaining a significant boost in response time. With high volume transactions in invoice and cash receipts processing, for example, this greatly improved processing efficiency.

The pressing challenges of our business entering our second year were hiring experienced, talented people and coding programs as fast as we could. In 1969, we hired our first employee, John von Jess. Von Jess was a big, burly man with talents for creativity in design and speed in programming. He would sit at the keypunch, programming on the fly, with cards fanned out between his teeth, fingers flying over the keypunch, deep in concentration, and steeped in the joy of the task. Von Jess, Charlie Silva, Chet Domoricki, and many of those to follow, were all experienced professionals we were able to hire away from IBM. We enticed them with the opportunity to be a part of a fast growing entrepreneurial company doing new things differently.

We were an unquestionable success by 1971, when ICP put out its first Million Dollar Awards for packaged software. We were one of 24 firms that were recognized that year for the first time in the history of the industry. We were generating a lot of revenue, but we were always just barely breaking even, as we were constantly called upon to plow revenues back into the business to maintain the rapid growth. While it was a good time for the software development business, cash flow was always our number one concern. American industry was entering a long stretch of high inflation and low productivity in the 1970s, the classic recipe for a cash crunch and recession. There was more than an occasional month when Watson and I both held back on cashing our own personal payroll checks to ensure that those we'd cut to our employees would clear the bank.

While it was an exciting time to be in business for yourself in the burgeoning independent software industry, it was also a fertile period in manufacturing inventory management. Groundbreaking effort in production management had been going on for some time, much of it spearheaded by countless IBM professionals working with manufacturers over the previous fifteen years. Many of these individuals had been recruited from customer sites for their demonstrated brilliance in working around the limitations of the first generation of electromechanical machines. Manufacturers themselves remained the test-bed proving grounds for much of the innovation that was beginning to rise to the fore in the discipline.

Pockets of innovation were active throughout New England, the Mid-Atlantic, and Midwest regions, with cross-pollination between regions, as individuals were courted and lured away from one Fortune 500 company to another. Pittsburgh, Cleveland, New Bedford, Racine, Worcester, Springfield, Minneapolis, and Baltimore were all fertile enclaves of early advances in the new "science" of production and inventory control. Joe Orlicky at J .I. Case, Jim Burlingame at Twin Disks, both in Racine, Wisconsin; Oliver Wight at Raybestos in Strafford, Connecticut, later at The Stanley Works, with George Plossl—these and countless others took to the field of applying computers to manufacturing management with a keen fervor. Many of these pioneers went on to join IBM, to work with people such as Paul Bacigalupo at American Bosch Amour in Springfield, Massachusetts; Ted Mussel, who worked with numerous IBM customers in Racine; and Gene Thomas, whose imprint was deeply etched in the original IBM bill of material processor program.

The most significant advance during this period was the development and spread of time-phased requirements planning. Time-phased requirements planning replaced the reorder point method for managing material stocking, availability, and commitment to orders. This advent was at the heart of what was to become universally known as material requirements planning, or MRP. MRP, a subset of the PICS model, was built around the bill of material processor, which calculated gross requirements, but went one step further. MRP addressed the critical question of "when" each item in the bill of material explosion was needed. Time-phased requirements planning had always been done on high-cost items, but manually. With the vast improvements in computing memory and processing speed that had been achieved in the 1960s, time-phased replenishment planning could be done cost-effectively using the computer on the entire bill of material. Time-phased requirements planning enabled efficient processing at the gross requirement, or end-item level, but also assigned a detailed schedule to the entire nested bill of material structure down to the lowest level. This moved material planning from its traditional base using historical data and the manual reorder point method of inventory stocking to a system grounded in projected future demand.

Dr. Joe Orlicky, at J. I. Case in the early 1960s, is generally credited as being the "father" of time-phased replenishment planning and hence, MRP. But even before he authored the seminal work on the topic in 1975, he estimated that there were perhaps as many as 150 MRP "systems" installed throughout industry by 1971. MRP was anything but a codified discipline during this time. Though Orlicky gets much of the credit, contributions were being made at almost every site that attempted an implementation, whether they succeeded or failed. It was such a new and untested field of inquiry that as much was learned from failure as success.

Exciting, yes. Wild, indeed. If ever there were a period warranting the warning caveat emptor, this was it. Gains were made, but losses suffered too, on both sides of the supplier/end user equation. And the situation was not destined to change, certainly not anytime soon. It was a time of hard lessons learned all around.

6 The Rise of the Church of MRP

Often intellectual or power elites hide their knowledge on purpose, to keep to themselves the advantages that go with the information. To accomplish this they develop arcane languages, mysterious symbols and secret codes that are meaningless to those not initiated into the guild.

Mihaly Csikszentmihalyi, *Creativity: Flow and the Psychology of Discovery and Invention*

To everyone who met him, Dr. Joe Orlicky appeared brilliant—or arrogant—usually both. Born to privilege in his native Czechoslovakia, he fled tyranny in his homeland after World War II and settled in the United States. Not known for his warmth or charisma, he was not without a droll sense of humor. He often told the story that the worst dilemma in fleeing his native country was whether to take his prized stallion or his girlfriend with him. The horse, he reported sadly, was left to the Communists.

Highly educated yet unable to command English with any great facility, Orlicky found it difficult to find work when he first arrived. He was unable to meet the rigors of the language for the jobs to which he was suited; yet was overqualified for placement as a menial laborer. He eventually found his way to Chicago, where he took an advanced degree at the University of Chicago. He subsequently found a job in the production control department of J. I. Case, manufacturer of heavy equipment in Racine, Wisconsin.

It was in Racine in the late 1950s that Orlicky began developing his ideas on computerization of inventory management. To the program for bill of material explosion, he added the calculation to determine requirement dates associated with each item in the bill. Time-phased replenishment planning was driven by the goal of matching material availability more closely with actual production schedules in order to develop the most cost-efficient means of meeting the production plan.

The fundamentals of MRP, as it was advanced at J. I. Case and elsewhere, were aimed at answering the simple questions: *what do I need; when do I need it; what do I have;* and *what must I order* to meet the production plan that management has given me? As Orlicky would later write in his seminal work, *Materials Requirement Planning* (1975) "to generate information for current order action is not the only function of an MRP system … but is the primary one."

MRP techniques, Orlicky wrote, "are expressly designed for dealing with dependent, discontinuous, non-uniform demand, which is characteristic for manufacturing environments."[19] The reorder point method was devised to maintain inventory based on some factor of historic use, economic order quantity, and safety stock. MRP did away with the tendency to apply the same ROP theory on dependent items. It did this via a combination of calculations processed against the bill of material, using the end-item due date coupled with manufacturing standard lead-time offsets to determine specific date requirements for each item at each level of the bill of material. This answered the question of when material needed to be available. As the sophistication of the program developed, MRP came also to address the other critical questions of *what do I have, what is available to use*, and subsequently, *what do I need to order?*

Fundamentally, the question is can I make what I need when I need to make it? The reports generated by processing an MRP calculation were vital to answering the second set of critical production control issues required to actively manage the flow of work orders in the plant, including: *can I increase or do I need to decrease a production quantity; do I advance the due date, defer, suspend, or cancel it?*

These questions, in total, represented the universe of issues an MRP calculation could answer. But the reality was far more complex than that. We all know what happens to the *best-laid plans* (of mice and men or manufacturing). This is nowhere, perhaps, more true than inside the four walls of the plant. But the potential advantage that the computer offered was significant for its ability to process the huge scope of data in these calculations more frequently, thereby approaching the true dynamic nature of events in manufacturing. A critical issue here, greatly relevant to MRP, is precisely that of frequency. In the early days of MRP—given the state-of-the-art of computer technology—this was not an insignificant question.

A full-blown MRP processing calculation was typically a lengthy run for any company with complex bills of material. A processing run could range anywhere from several hours to several days. Due to competing demands for computer time, with manufacturing typically last on the list of corporate priorities, an MRP run was done either at night or on the weekend. This fact—coupled with the very dynamic nature of events in the plant, with new orders coming in, parts proving defective, machines breaking down—meant that the material plan was always, to greater or lesser degree, out of sync with the reality on the shop floor.

The solution to this combination of circumstances was "net-change MRP." In his book, Orlicky writes that "inherent to schedule regeneration, always a big job, is the task of massive data handling which entails a delay in obtaining the results of the requirements planning run and dictates that the job be done periodically, i.e., at economically reasonable intervals. This causes the system to be out of date, in some degree, at all times. How serious a disadvantage this represents in a given case depends on: the environment in which the MRP system must operate; and the uses to which it is being put."[20]

He writes that "there are two basic alternatives of MRP system implementation: 1) schedule regeneration, and 2) net change. The first of these affords high data processing efficiency but limits the frequency of replanning, as a practical matter to a weekly or longer cycle. The second is designed for high frequency (or continuous) replanning, at the expense of overall processing efficiency." He continues that "net change is less efficient, and therefore more costly, primarily due to multiple access to inventory records in transaction posting" But, "in net change material requirements planning, the emphasis is *on inventory management and production planning efficiency*, not on *data proceessing efficiency*."[22]

Net-change MRP, in essence, enabled greater frequency of updating the plan because only those requirements that were affected by changes to the plan or events in the factory were processed. This was an advance that was made possible by random access technology that became available in disk-based systems, such as the Ramac 305, the 1400 Series, and the System/360, which would have been impossible to achieve with tape-based systems, where an entire tape would have had to been read, processed, and reread repeatedly.

Though Orlicky is widely credited with initiating the first net-change MRP processing run in 1961, it had been performed elsewhere before that. Paul Bacigalupo, an IBM systems engineer working with American Bosch Armor in Springfield, Massachusetts, engineered a net-change calculation for his client in 1959, employing the IBM Ramac 305 disk-based computer. Bacigalupo doesn't claim that his was the first ever, for he credits the basis of his work on a case application note written up on an earlier IBM installation. Though Orlicky labeled it and popularized it, net-change calculations seem most certainly to have been engineered previously elsewhere.

Time-phased replenishment planning and net-change MRP were radical advances for the time, for adding real demand to the calculation of material requirements. These two technical achievements, as much as anything, transformed the practice of production and inventory management in American manufacturing in the second half of the 20th century. What had once been viewed as a lowly calling was becoming a field of increasing professional interest.

Out of the grassroots of the discipline, the American Production & Inventory Control Society (APICS) was created in the spring of 1957. Twenty individuals met in Cleveland to discuss the idea of elevating the stature, if not the practice, of production and inventory control by means of a professional association. The purpose of the society was to improve the standing of people involved in this critical function in American industry, for without material that is efficiently procured and managed, how can anything be made cost-effectively? Central to the precept of the organization, in addition to merely finding strength in numbers, was the principle of training and education. The regional structure with local community chapters was settled upon, and New Bedford, Massachusetts, became the first chapter formed that year, with a membership of 126 practitioners.

Monthly local chapter meetings and an annual national conference were the primary venues for participation during the early years. IBM donated $5,000 to the organization in 1959 in order to fund the position of an administrative assistant. In 1962, membership stood at 2,300; in 1965, it had more than doubled, to 5,000 members.[23] Member recruitment was primarily word of mouth. Chapters spread from one community to another as word spread and as practitioners relocated to take up new jobs around the country. Dr. Joe Orlicky was instrumental in the founding of the Minneapolis chapter. Oliver Wight, a young production and inventory control manager at Raybestos Corp. in Strattford, Connecticut, was involved in starting the Fairfield County, Connecticut, chapter in 1959. Wight had graduated from Northeast College in New Hampshire with a degree in English, and when he returned home to Strattford to find work, he was hired by Raybestos and assigned to the inventory control department, deemed by the personnel manager at the time as the most appropriate match for someone with a degree in English.[24] Wight took to the field with enthusiasm.

The organization's strength lay in the local chapters through much of the 1960s. The local chapters managed to put together and host the annual conference year to year, which served as the main source of financing for the national organization. Little was done in the way of development of a full-fledged education and certification program, however, due to the tenuous nature of financing. Though the first executive director was hired in 1968 and national offices opened in the Watergate complex in Washington, D.C. (the floor adjacent to the Democratic National Committee offices in the complex), the organization closed out the decade in dire financial straits. Membership stood at 8,000.

Joe Orlicky had moved on from J. I. Case to IBM, where he was recruited in 1962. His role was to educate senior executives at IBM client sites to the benefits of applying computer technology to managing inventory and production control. One of the people he encountered in his work as an educator and promoter of the emerging discipline was Oliver Wight, who was then working at The Stanley Works in New Britain, Connecticut. Orlicky was attracted to this bright and gregarious young man by his passion for the field. Wight joined IBM in 1965, where he remained until 1968, when he left the company to work with George Plossl, with whom Wight had worked at The Stanley Works. Plossl and Wight operated briefly as an education and consulting firm in the industry, until they separated to pursue individual paths in 1969. But their partnership left an indelible imprint on the industry through their co-authoring of *Production and Inventory Control: Principles and Techniques*. This book became something of a bible to production and inventory control practitioners, industry analysts, consultants, and anyone otherwise interested in the subject. It was as though Plossl and Wight had inscribed their ideas in stone for the sanctity with which the book was treated. Where it clarified and codified many critical principles and techniques at its first printing, it had the effect of barring the door to introduction of new ideas that were even vaguely contradictory for the next twenty years. In the age of information technology, twenty years represents time in epochal proportions. But in the area of manufacturing control systems, if Plossl and Wight had not blessed an idea, it was suspect from the beginning.

APICS was never more than lukewarm to outside consultants in the 1960s. The concern was that consultants were primarily interested in selling services, rather than truly promoting the disciplined practice of production and inventory control. This changed abruptly, however, in the early 1970s, as the financial circumstances of the organization grew more pressing. In 1972, at a meeting attended by veteran professionals of both the industry and APICS, a plan was formed to bring the growing consulting wing of the industry more closely together with the national organization in a cooperative venture. The aim was threefold: to improve the financial underpinning of the APICS organization, to initiate a series of national seminars and training sessions that would inaugurate the long-desired professional APICS certification program, and generally to promote the critical importance of the modern practice of computer-based production inventory and control management.

The campaign became known as the "MRP Crusade." It featured a host of seminars taught by industry consultants, where attendance was charged, but the services of the half-dozen or so consultants were donated. The seminars were coordinated and promoted by the local chapters, who received, in turn, a 50–50 split of the revenues with the national organization. Wight, Plossl, Jim Burlingame, Walt Goddard, and others conducted the series of seminars. Joe Orlicky got IBM to contribute financial backing to sponsor recordings of the events so that the training material would be more widely disseminated.[25]

This groundswell of activity in 1972–73 led to creation of the APICS certification program in 1973. The subject of the first two classes in the Certified Production and Inventory Management (CPIM) regime focused on forecasting and inventory planning. These activities also served to greatly boost membership in the national organization. Membership rose from 8,000 members in 1970 to 12,000 in 1975, and went up over 27,000 members by the end of 1978. By the end of the decade, the organization counted 46,000 members.[26]

If the 1970s were good years for APICS, they were golden years for Oliver Wight and what was to become an extensive consulting, software evaluation, and education powerhouse. Wight's enterprises went through an array of permutations during this time, but all were sparked to success by the charismatic charm that inspired just about everybody with whom he came in contact.

Oliver Wight was gregarious and outgoing and was gifted with the flair of a raconteur. Known as "Red" by intimates, and simply Ollie by everyone else, Wight was red-haired, good looking, and broad shouldered, and rarely, if ever, did he meet a fellow he couldn't warm with his easy manner and quick smile. Wight commanded a room as did few others, and before an audience of a half-dozen or several thousand, he knew how to capture the interest and generate a sense of shared commonality such that it was always a pleasure to be in his company. Larger than life, he was human, too, and though he loved to make money, he also spent it freely. He was legendary in his extravagance in the collection of fine automobiles as well as for his generosity in tipping all the service people at every hotel he ever visited. He was, in short, perfectly suited for what was to become the mission of his life: the fostering and promotion of the fledging MRP industry into something few others could even imagine.

Wight's core enterprise was centered on education. He spoke knowingly from firsthand experience of the merit and value of using computers to help run manufacturing. His experience at Raybestos and The Stanley Works, and his opportunity to work with a wide variety of other companies while at IBM provided validation to all his claims. More importantly, he spoke with heart-felt zeal that though computers were marvelous contraptions, they were nothing unless people knew how to use them. (Wight dismissively referred to computers as the "C" item before large crowds of inventory management practitioners, who understood the reference to their system of labeling material by an "A-B-C" priority system—with "C" having the lowest priority). And beyond the mission of promoting their use, Oliver Wight was expert at capitalizing on the rapid expansion of opportunity to help others learn how to use them to the fullest benefit.

Oliver Wight—with Walt Goddard initially, and then a whole host of individuals whose careers would become synonymous with MRP and, later, with MRP II—offered educational courses to senior management and middle-level, operational management as well. In the early days of Oliver Wight, Inc., he offered a two-day top-management course as well as a five-day middle-management course.

When Bill Watson and I left IBM to start Software International we asked Wight to serve on the board of directors. We were friends and shared a common belief in the potential of the industry. He generously supported what Watson and I were working to accomplish in the early years of our company.

In 1971, however, he came to me and announced he was resigning from the board. He had to, he told me, for it would be a conflict of interest in a new venture he was undertaking at Wight and Associates. At this time, there were perhaps 30 or 40 "packaged" MRP solutions on the market. As he and I both knew, there was a lot of "smoke and mirrors" and hyperbole around much of what was being promoted. Wight had decided the time had come for someone to step forward and perform industrial-strength "consumer report" evaluations to the functional claims of the various packages. The timing was exactly right, I thought, for there was very little consensus for what constituted an MRP system "under the hood." There was little commonality in terms and almost no way to adequately compare one system with another. If Wight was to do the job that needed to be done, he said, he needed to be independent of affiliation with any particular vendor.

Wight wrote in the introduction to his "MRP Software Evaluation" guide that the evaluations were meant to "get companies through the design and evaluation stage quickly and into implementation." Companies were currently finding these steps to be huge stumbling blocks, being "expensive and time consuming." The Oliver Wight group, he stated, would be "able to do an evaluation in greater depth and from the vantage point of considerable experience ... [where the] cost can be distributed over many companies."[27] The cost to the end user was not insignificant: Wight offered a notebook defining the general framework for the "standard" MRP system, priced at $675; his evaluations of individual packages were typically priced upwards from there by roughly a factor of ten.

Though modest in scope, in that Wight only targeted evaluation of system functionality, not performance or cost-benefit, the MRP Software Evaluation service he established had far-reaching and profound impact on the industry. At the heart of his evaluation was a comparison of a vendor's functionality against what Wight termed a "standard" MRP system. "Evaluations ... referred to a standard or complete system," he wrote in the introduction to the volume. "The standard is not an ideal system with all the possible functions that could exist in a system. Instead, the standard system is a simple comprehensive set of tools."

Further, he wrote, the "definition of what constitutes a standard system is not at all arbitrary," that experience "demonstrates that the fundamentals of planning, scheduling, and coordinating all the different functions within a manufacturing system are the same from company to company. This experience confirms that a standard set of tools ... applies as well to a company making brassieres as it does companies making jet engines."[28] Though he qualified that the definition of what was standard was an opinion, he also claimed that no one else had the level of experience that his group had. With Wight's growing prestige in the industry, his claim as to what was standard was supremely regarded.

For the time, the benefit was significant, for it brought a semblance of order and discipline to the hugely uncorralled, rapidly expanding manufacturing software industry. It provided a framework, or target, that many of us developing software were forced to work toward. The problem was that, ultimately, there was no such thing as a standard system. And his view, despite his claim that "there is not a lot of room for opinion on what will work and what won't work"[29] was still only an opinion.

Though an important event at the time, given the then state-of-the-art of things, Oliver Wight's standard system evaluations had the effect of casting in stone the specifics of what constituted MRP. APICS adopted Wight's standard model as its model for education and certification of practitioners, ensuring that anyone who worked in the field as a practitioner or consultant followed the Wight model to the letter of the law. This had grave ramifications, some which would come back to haunt not only later efforts on my part, but, coupled with the required conceptual "work arounds" engineered in the original PICS manual I helped create in the early 1960s (including standard lead times, capacity planning, level-by-level, and exponential smoothing), they would haunt a whole segment of industry, if indeed, not the entire industry, as we later moved closer to the 21st century.

The 1401 Card Reader and 1403 (courtesy of IBM). The 1401 Card Reader read 600 cards per minute. The 1403 Printer printed 1000 lines per minute. The Central Processing Unit (rear center) contained up to 16k of memory; if more memory was needed, another box had to be added.

Notice that there is no keyboard, typewriter, or any other form of input device besides the card reader and the switches on the CPU. When the program hung up (bugs, of course), the operator read and recorded the memory position and then presed the Print Memory button, which dumped the entire contents of memory onto the printer. Knowing the Core Dump and Hang-up Position, the programmer could then proceed to debug the program step by step.

The 1403 Printer was a marvel for its time. Nothing approaching its speed had ever been seen. The control of printing page length was a punched endless paper tape that resided in the upper right corner of the printer. If the tape broke, an entire box of paper would be thrashed through the printer before the operator could push the Stop button. Worse yet, when the printer sensed the end of the paper, the top of the printer would automatically rise without warning, thus spilling any objects, solid or liquid (coffee, perhaps), all over the printed paper at the foot of the machine.

There are two tape drives at the right in the picture above. The tape commands were % UnR and % UnW, where n was either Drive 1 or Drive 2. Operators had to program their own tape error routines, which were usually something like this:

```
Error 1.  Backspace #1
          Add 1 to CTR
          % U1R (read #1)
          IS Counter = 10
          Go to ERROR
          Halt #1
```

This routine simply tried 10 times, then if still failing stopped the tape drive.

The 1405 RAMAC (Random Access Method of Accounting), circa 1960 (courtesy of IBM Archives). This machine held either 5 or 10 mB of characters (prior to bytes). This model was 5 mB. Note the arm mechanism on the left; this arm moved up or down, just like the old Wurlitzer jukebox player. Only the one arm serviced all platters. Each record was 100 characters long, with ten records per each track. In order to program the 1405 RAMAC, the operator gave a Seek command. This command moved the arm, then a Read or Write command actually read or wrote the data. The predecessor to the 1405 was the 305 RAMAC, which looked about the same as the 1405.

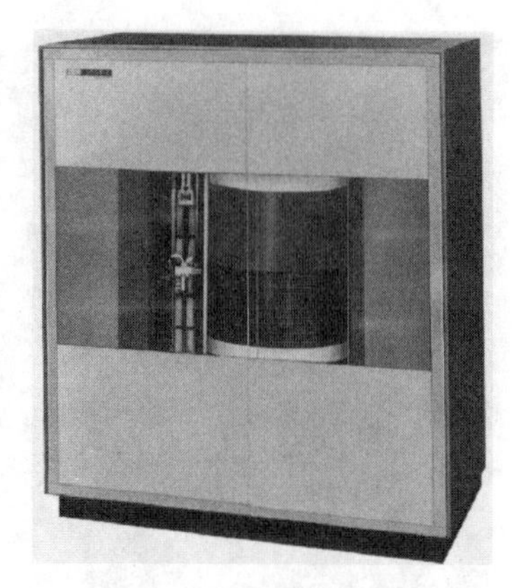

The 1410 (courtesy of IBM). The 1410 was the big brother of the 1401; it was faster and had more storage. This model has three 1411 disk drives with removable disks, each disk containing 3.75 mB. The massive cabinets in the rear were memory boxes. These systems cost between one and five million dollars. One could also buy a 50 megabyte hard disk at an exorbitant price.

Again, there is no keyboard or typewriter input for the 1410, only cards. There were two of these computers in Worcester County in 1962. At one location, the first application was the bowling league standings for the company.

The System/360 (courtesy of IBM). The Model 30 had three 2311 disk drives, each with 7.5 megabytes. The term bytes as well as hexadecimal arrived with the 360.

Also, note the IBM Selectric typewriter, a system prerequisite, but never used by a good programmer because it would cause the computer to prompt and stop. Users never wanted this machine to stop until the job stream was finished!

This configuration cost over a million dollars in 1964 (equal to $10 to 20 million in 1999 dollars). This system was always justified by, purchased by, and implemented for finance. Manufacturing applications, while desirable, were always the orphaned cousins to financial requirements.

Paul Bacigalupo, George Plossl, Joe Orlicky, and Bill Jones, circa 1974. Paul was APICS President in 1974. George and Oliver Wight wrote the definitive book, *Production and Inventory Control*. Joe was the "Father of MRP," as proclaimed by IBM.

ProfitKey celebration, 1986. From left, Dick Lilly, Rich Lagoy, Bob Davis, Frank Maglio, and Dave Layne.

Ron Ripley, Dick Lilly; Lilly Software's 1999 Customer of the Year, Loyal Peterman of Abrasive Technology, Dave Layne, and Tony Maurno at the Lilly Software User Conference, Orlando, Florida.

US005787000A

United States Patent [19]

Lilly et al.

[11] **Patent Number: 5,787,000**

[45] **Date of Patent: Jul. 28, 1998**

[54] **METHOD AND APPARATUS FOR SCHEDULING WORK ORDERS IN A MANUFACTURING PROCESS**

[75] Inventors: **Richard T. Lilly**, Hampton Falls; **David V. Layne**, Litchfield, both of N.H.

[73] Assignee: **Lilly Software Associates, Inc.**, Hampton, N.H.

[21] Appl. No.: **250,179**

[22] Filed: **May 27, 1994**

[51] **Int. Cl.**[6] **G06F 19/00**; G06G 7/64; G06G 7/65

[52] **U.S. Cl.** **364/468.01**; 364/468.05; 364/468.06

[58] **Field of Search** 364/468, 478, 364/401, 402, 403, 468.02, 468.03, 468.05, 468.06, 468.07, 468.13, 468.18; 395/228, 229; 705/8

[56] **References Cited**

U.S. PATENT DOCUMENTS

4,459,663	7/1984	Dye	705/29
4,646,238	2/1987	Carlson, Jr. et al.	364/403
5,089,970	2/1992	Lee et al.	364/468.02
5,101,352	3/1992	Rembert	364/468
5,216,593	6/1993	Dietrich et al.	364/402
5,233,533	8/1993	Edstrom et al.	364/468
5,283,745	2/1994	Tanaka	364/468.05
5,291,397	3/1994	Powell	364/402
5,303,144	4/1994	Kawashima et al.	705/8
5,325,304	6/1994	Aoki	364/468.06
5,440,480	8/1995	Costanza	364/401
5,463,555	10/1995	Ward et al.	364/468
5,586,021	12/1996	Fargher et al.	364/468.06
5,630,070	5/1997	Dietrich et al.	705/8

Primary Examiner—Joseph Ruggiero
Assistant Examiner—Sheela S. Rao
Attorney, Agent, or Firm—Testa, Hurwitz & Thibeault LLP

[57] **ABSTRACT**

A computerized system is provided for scheduling a plurality of work orders in a manufacturing process. Each work order to be scheduled specifies a set of operations to be performed using a plurality of resources and materials. Data including resource availability information for each resource used in the manufacturing process, material availability information for each material used in the manufacturing process, and work order information is received and stored in a computer. The work order information includes a release date for the work order, a want date for the work order, operations information, and material requirements information. The operations information includes the identity and sequence of operations to be performed for the work order, the identity of the resources needed to perform each operation, a minimum resource capacity needed to perform each operation, and the time needed to perform the operation. The materials information includes the identity of the materials needed to perform each operation and the quantity of each material needed for the operation. Resource capacity and a start date/time and a finish date/time are assigned to each operation based upon the resource availability information, the material availability information, and the work order information. The assigned resource capacity, the assigned start date/time, and the assigned finish date/time for each operation are displayed on a computer screen in a graphical format.

23 Claims, 11 Drawing Sheets

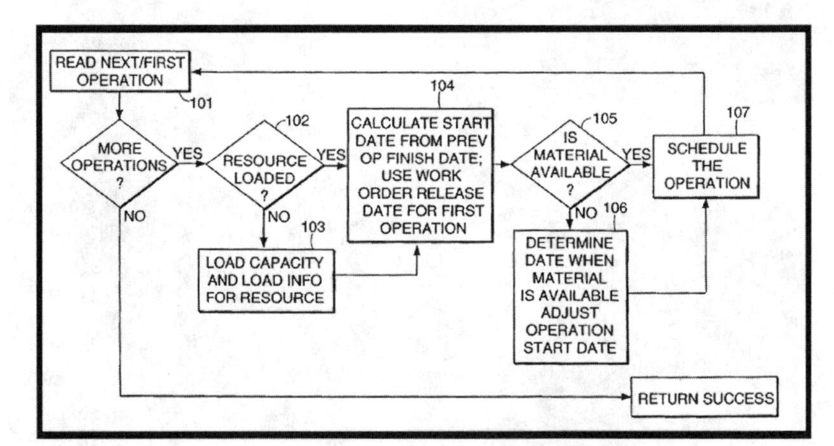

Dave Layne and I were awarded United States patent number 5,787,000 in July 1998. The patent was granted to us for our software design for concurrent scheduling of material requirements and operations.

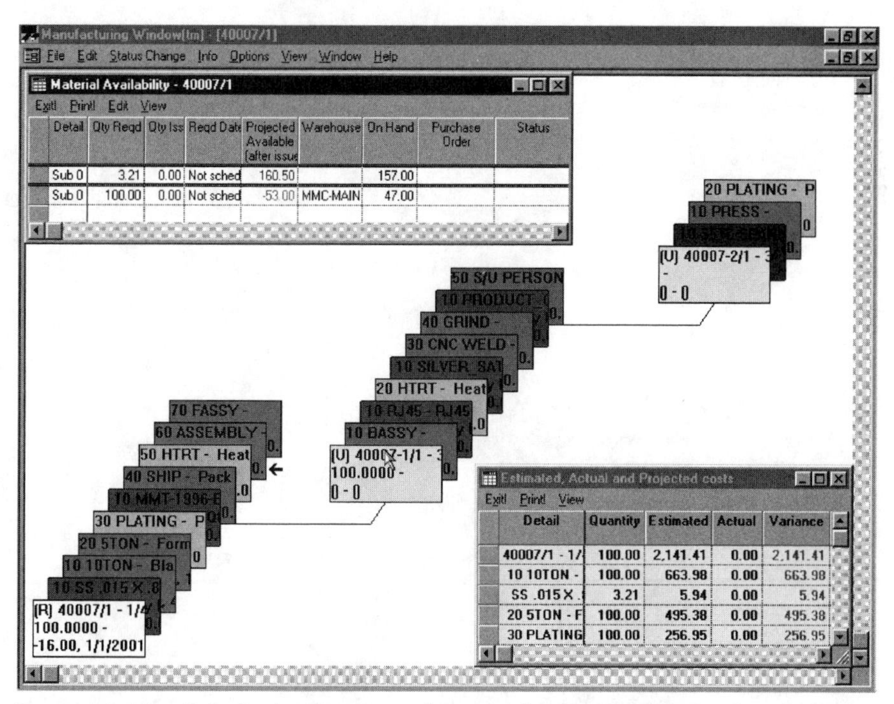

Detail	Qty Reqd	Qty Iss	Reqd Date	Projected Available (after issue	Warehouse	On Hand	Purchase Order	Status
Sub 0	3.21	0.00	Not sched	160.50		157.00		
Sub 0	100.00	0.00	Not sched	-53.00	MMC-MAIN	47.00		

Detail	Quantity	Estimated	Actual	Variance
40007/1 - 1/	100.00	2,141.41	0.00	2,141.41
10 10TON -	100.00	663.98	0.00	663.98
SS .015 X .	3.21	5.94	0.00	5.94
20 5TON - F	100.00	495.38	0.00	495.38
30 PLATING	100.00	256.95	0.00	256.95

The graphical "card" display in VISUAL Manufacturing. The Manufacturing Window shows all of the operations, materials, and services for a work order.

SECTION II: BREAKING THE MOLD

Breaking Ground

"In computing, we don't do any real work. We just transform one pattern into another."

Ralph Gomory, IBM Research Director

Software International was successful after only its first few years in business. We weren't the biggest software company, nor even the biggest in our market niche, but we were making money, continuously hiring more staff, building out our software, growing our customer list, and polishing our credentials as one of only a handful of frequently short-listed MRP application vendors. And I was restless.

More than restless; agitated. I was spending more and more time managing the activities and concerns of a growing company, rather than working directly with customers and writing code. But my discontent was deeper than that. Though I was looked upon by my peers as successful, I was sinking into a classic mid-life funk, wondering *is this all there is*?

We had lost two children to a mysterious malady. Richard T. Lilly III, whom we called "Timmy," had died in 1964 at age six; and Shawn Andrew in 1969 when he was seven. Their illnesses went undiagnosed while we struggled to care and comfort them. (It was subsequently determined that they had died of Reye's Syndrome.) During this period, I was drinking heavily and there was stress in my marriage. Growing up as the first-born son, I had always labored to please others, my parents—my father, especially. I was used to setting the bar impossibly high, and just as used to finding little praise or encouragement—or satisfaction—in clearing it. As an adult and a parent, I expected a lot from others, perhaps too much, especially from my family, but never more than I expected from myself. I was a hard task master, but never so hard as on myself. But after getting Software International up on its feet, what with the circumstances that surrounded my life at the time, I felt a deep emptiness. There was little joy or comfort in my work, nor in my life.

I felt as if there were nowhere to turn.

Surrender is hard. Especially for someone used to being in charge. And yet, faced with the impossible, irresolvable angst that filled me, surrender was the only option that seemed open. During this period, I was moved to turn in a direction I'd never before seriously considered and was shook to the very foundation of my existence by a spiritual conversion that turned my life inside out. Putting my faith on the line in putting my life in the hands of God was the most difficult thing that I had ever done. But the experience of surrender was transformative. Immediately. I left work and went home to talk with my wife. She listened, but I could tell she was skeptical. She knew me only too well. But in time she came to believe in the transformation of my experience, if for no other reason than from that day forward, I never took another drink of hard spirits.

Trying to run a company when two partners both want to be the boss is a difficult challenge. Bill Watson and I made a good team, but we were both headstrong. And though I had won the toss of the coin to become president, the arrangement was not without its friction. When Watson came into my office one day in 1974 to tell me he wanted to talk about the possibility of buying me out, I surprised him. "Let's talk," I said. And so we did.

Our settlement was straightforward, one that benefited us both. We arranged for him to pay out a three-year salary package over five years, beginning in January 1975. Though I would remain a member of the board, I would forgo all authority in the day-to-day operations of the company.

At this time, we were living in Groton, Massachusetts, where my son Michael was enrolled as a day student at The Groton School. My daughter Suzanne was in middle school, and my youngest son, Mark, was in elementary school. We were building a large addition to our house, and life-after-Software International for me began with completing the finish work on the new space.

Free of the yolk of managing a company, I was still restless and casting about for what I might do with the rest of my life. I dabbled some in real estate. I love history, particularly military history, and thought about writing a book. I'd always been fascinated by the rich complexities of Benedict Arnold—bold, audacious, overlooked, and under appreciated as a great leader of men and as a military tactician—and I began preliminary work on a novel based on his life. Never one short for an opinion on anything, I also thought about running for political office. Born a Democrat, I was dismayed at the state of affairs of both the economy and national politics with Jimmy Carter, a Democrat, in office. We seemed drifting as a nation: our industrial base was deteriorating, we were getting clobbered by foreign competition, and inflation was scouring out our sense of financial security. I briefly explored a political candidacy, to the point of meeting with political consultants and potential advisors. In the end, I decided I just didn't have the heart—or the stomach—to be a politician.

Midway through my five-year hiatus, I decided what I needed was a real change, a change of locale. I was more enthused than everyone else in the family that it was the right thing to do. By then, Michael was off to college at the University of Pennsylvania. My daughter Suzanne was a junior in high school and understandably opposed to moving. Mark, my youngest son, in the middle of his eighth-grade year, saw the move as something of an adventure.

We left Massachusetts on New Year's Eve day, 1977, following far behind the migrating snow geese for the inviting warmth of coastal Florida. When we arrived in Marathon, midway down the Keys a week later and read about the big blizzard that had buried New England in our wake, I thought it provident, a good omen for our move. The days were sunny and warm—shirt-sleeve weather in the middle of January, the nights mild and clear. I looked to the future as being about as wide open as the vast horizon of endless sea that engulfed us.

The Florida Keys are literally "the end of the road," the farthest south you can go in the contiguous United States. First sighted by Ponce de Leon in 1513, he called them *Los Martires*, the Martyrs, for their tortured, twisted look. Home to hurricanes, shipwrecks, and pirates, they have always been a haven from the staid life and have drawn vagabonds, outcasts, and lovers of the sun and a simpler life almost from the time they were first shown on a map.

I am an avid sports fisherman and we had vacationed in the Keys on several occasions. Built on the bedrock of ancient coral reefs and limestone beds that took their original form some 4,000 years ago, the Keys are home waters to over 600 species of fish. Here you can fish for Blue Marlin and the Great Barracuda. Or you can pursue more exotic fair, including bonefish, permit, and tarpon. The coral reefs that are home to many of these fish began to form in these waters 500 million years ago. Coral needs light and warm, clean waters to flourish. In a manner of speaking, it was exactly the same mix of ingredients that I was seeking in moving my family there in 1978. I had no real plan or agenda to accomplish other than to live and let live and be left alone—at least for a while.

We settled in Marathon, a small community forty miles north of Key West. Marathon was little more than a wide spot that straddled the desolate two-lane stretch of U.S. 1 that ran to the tip of the Keys. It had been originally settled in the early 1800s by New England fisherman who came south during the winters to fish. It was briefly known as "Conch Town" in the middle 1800s, when settled by Bahamian nationals who were drawn to the lucrative trade of salvaging shipwrecks. It was named Marathon in the early 1900s, when it served as the construction base for building what was then known as Flagler's Overseas Railway, a line of track to carry vacationers from Miami all the way to Key West. One of the hired track hands beknighted the place with the offhand remark that laying track through the Keys was a displeasurable "marathon." The railroad came and went, but the name stuck.

I had not come to the Keys to contemplate the meaning of life, to spend my time idling in the shoals or combing the beaches. I am happiest when engaged in something that requires me to be creative, to rely on the craft of my own daring, to envision and pursue a plan with earnest commitment until the thing is done. The thing I had set before me was to build a house. To lay the foundation and erect the timbers, to raise the roof and set the threshold; to do this with my own two hands. I had the luxury of time to do this, but was aware, too, that time on my five-year severance from Software International was nonetheless slowly ticking away.

Building a house is not unlike building a software program, though no doubt leagues more tangible and concrete. You start by asking very broad questions as to the purpose and function, how do the various functions fit together, and what will it "look" like. You start with a general idea and ask progressively more refined questions as you go. Eventually, you lay out a spec, or blueprint; you assemble the necessary resources; and you break ground. With mounting anticipation, you look ever increasingly to the moment you can "move in" and inhabit the space.

My wife Laura and I drew up the rough blueprint together, and we had a draftsman complete the detailed plan. We started with knowing what we didn't want, which was an ordinary block of a house. Because we were building in a hurricane zone, the primary living space had to be elevated above the potential surge line. At ground level, therefore, we had room for an enclosed utility space and open carport. The main part of the house sat atop that. It was to be modest, yet spacious, with four bedrooms, two baths, a living room and dining room, and a large patio deck overlooking the bay in front of the house. The exterior was to be done all in cypress board-and-batten style construction. To secure the exterior, we would have to use brass nails, as the salt air would eat the heads off anodized nails in only a year or two. I remember that the brass nails represented a small treasure, costing a nickel apiece at the time.

We broke ground in April. My goal was to be in by Thanksgiving. With my oldest son, Michael, off to college and my other two children settling into new lives in new schools in Marathon, every morning I would walk from the house we rented on Sombrero Beach around the corner to our lot in an elbow of land adjacent to Tingler Island. I wore comfortable work clothes and carried a lunch cooler. At the site by seven every morning, I would put the lunch cooler down somewhere in the shade and pick up my hand tools to put my vision of creating a physical thing to the test.

It felt good to labor under the bright Florida sun. The days were warm, building toward the swelter of summer, as the structure took form. I worked primarily alone, though occasionally I would hire a high school student to help me with some particular difficult task requiring another set of hands. Michael and a friend, Rick Diamond, came home from college to help that summer. Suzanne and Mark were enlisted as well. The roof was on by August and Michael and his friend suffered the dirty job of insulating the house at the height of summer. Suzanne and Mark spent many an hour sanding the 8" mahogany plank floors that ran throughout the house, down on their hands and knees with belt sanders, required to take as much as an eighth of an inch or more off each irregularly planed board imported from South America. It was a memorable time for us all, and we look back on it with fondness.

An acquaintance at the Sombrero Country Club, where I infrequently played golf, asked in passing how the house was coming. He had one under construction as well, only a professional builder he had hired was doing his. When I said we were shooting to be in by Thanksgiving, he laughed. He had broken ground on his house about the same time we had, and he wasn't going to be in, he said, until Christmas. Well, we'd see, I said casually, disarming his confidence. But I was determined to make the Thanksgiving deadline.

The interior of the house had something of a Moorish feel with clean, white stucco walls, high-pitched ceilings, and rounded arches for all the doors. The rooms were all designed to be as open as possible to catch and carry any breeze there was, yet with the ability to be closed off as desired. Every room had a ceiling fan for circulating the air. The living room, which looked eastward out on the bay, was constructed such that stereo, television, and bookcases were all recessed, giving the space an elegant, cleanliness. The interior doors and their casings were all made from mahogany-crotch lumber, which was beautiful when finished, but a bear to work with, especially in doing the detailed finish work around the frame. The kitchen cabinets were done in teak veneer, which had to be carefully sized and cut to get the proper look and durability to the design. In all, it was an entirely hand-built house, built with the hands of family, excepting the drywall and tile work, which I contracted out.

Our first Thanksgiving in Marathon was a special time of grace, seated at the family table in our new home. Everybody was there. We gave thanks for our blessings and good fortune. Competitive as I am, I couldn't help but feel a sense of pride in accomplishment when I found out later my acquaintance at the country club managed to get in by Christmas—Christmas the following year. But there was little room for gloating. The house built, the end of my five-year severance from Software International rapidly approaching, it was imperative I turn my mind to what I was going to do next to support my family.

I had no clear idea. I was, in fact, wildly open-ended in considering prospects. I considered starting a small business of some kind in Marathon, perhaps going into real estate. I even considered the possibility of taking on a franchise, such as a convenience store—anything that would support our life at the end of the road in the Florida Keys.

My son Michael had returned to Marathon at the request of the university to take a year off due to lacking scholastic performance. Certainly no dummy, having graduated from Groton with honors, he simply wasn't paying attention to the task at hand. When he lost his job at a local resort for some intemperate stunt, my wife and I were concerned whether he was going to find his bearings before foolishness took serious hold.

Perhaps, I told Laura, I might start a small software business, something Michael and I could do together. It was something I knew. It was something where I would be creative again, building something from little more than an idea in my head. And it was something I might be able to pass along to Michael. The more I thought about it, the better it sounded. I'd started from scratch before and done all right, and had no reason to believe we couldn't do it again.

8 Starting Anew

Technology is about making things predictable and repeatable so we do not need to devote so much time and attention to them.

Tor Norretranders, *The User Illusion*

Michael and I set up shop in the first floor of the house in Marathon. We installed a couple of Radio Shack TRS-80 personal computers with floppy disk drives. "State of the art" for the time, each double-sided diskette was capable of holding a half megabyte of data. Michael had done a little programming in Focal in high school, accessing a Digital Equipment PDP8 machine, but he'd never worked in BASIK. I handed him a BASIK manual and told him to read it while I went out to solicit business.

We signed our first contract for a custom program in the spring of 1979 with the largest real estate company in Marathon. Our charter was to develop a software package for automating the task of calculating a property's value based on a survey of its features. They fronted $5,000 to scope and develop the software. The most challenging part of the job was getting it to fit within the memory constraints of the hardware. We delivered what they needed, but kept the rights to the software. We didn't get rich, but we did manage to sell about a dozen copies of the same program in the ensuing months. We named our company simply Key Systems.

We were a classic job shop—programmers for hire, willing to build whatever application you wanted or that we could persuade you that you needed. We did a bit of everything, from job costing for a New England manufacturer to hotel and apartment management packages, to a point-of-sale register for the bar at the local yacht club, as well as a special program for a north-Florida physician involved in clinical drug studies.

It was in the early months of Key Systems that we met Dave Layne. Dave was working for his stepfather building custom homes. We happened to meet as a result of an offhand comment I had made to a woman I'd met at church about some difficulty I was having with my computer. She told me to call her grandson who, she claimed, could fix anything. Taking her at her word, I gave him a call. Dave came by and asked what the problem was. He had never seen a TRS80, but I showed him the error message that kept flashing on screen about the motor speed being too slow. Dave took the housing off, studied the machine's innards awhile, made a few adjustments to a potentiometer, and that was it. It was fixed. When it happened a second time some weeks later, I called him up and asked him to come over and work his magic again. The second visit, he became more curious about what we were doing and began coming around on a regular basis after that. I learned that he was interested in computers, had spent some time while he was recently in Maryland availing himself of the generosity of a local computer store. He said he'd read just about everything on their bookshelf while hanging around the store and had gained sufficient time on the display models to teach himself the rudiments of the technology. I asked if he would translate a payroll system we'd found in the public domain into BASIC for a customer of ours. I showed him the manual, and after pursuing it for a few minutes, he said he'd give it a try. His only caveat was that we furnish him a computer to work on at home. I gave him a machine and agreed to pay him $500. It was quite a job before he was through, Wang BASIC being considerably more robust that Microsoft BASIC at the time, but he managed it with real finesse. I contracted with him to do several other projects for us. He took to it with amazing ease.

Key Systems was gaining a reputation for the ability to program just about anything that anyone needed. Michael and I were busy enough that I began courting Dave to come to work for us. He was happy the way things were, working for his stepfather during the day, freelancing for us in the evenings and on weekends. In the fall of 1980, business had picked up enough that I made him a formal offer. He was reluctant to accept, afraid he would be letting his stepfather down. I let things go on the way they were for several months, letting him know the offer still stood, but not wanting to pressure him into a decision. Around Christmas time, he came by and told me he'd made up his mind. He came to work for us full-time in January 1980.

We rented a store front soon after in a Winn-Dixie shopping center on U.S. 1. We began staffing up with other programmers—John Houck, Clay Black, and others—talented people who seemed to appear as if out of the woodwork, out of the tropical thatch palm hammocks and mangrove thickets of the Florida Keys. Houck had run his own computer business in North Carolina and was looking for a change. Black was a former Rand Corporation analyst, a teller of tales of covert operations in the Far East, now enjoying the life of a recluse on a sailboat in the Florida Keys. Situated between a lounge and a Chinese restaurant, our office was rather nondescript but our sign was an obvious curiosity, at least to anyone who knew anything about computers.

One morning, a deeply tanned, longhaired beach bum in cutoffs and flip-flops wandered in off the street, curious about what it was Key Systems did. Dave Layne chatted with him and was impressed enough to encourage me to talk with him. Hair halfway down his back, our visitor didn't make a very positive impression at first glance. I was reluctant, but Dave wanted to give him a chance. Without telling him what it was, I handed the young man an accounts payable program I had written. I told him to take it and look it over and when he could tell me what it was, to come back and see me. I thought that was pretty much the end of it.

Before the end of the day, he was back in the building. He handed me the program, told me what it was, and then went about telling me where all the bugs in the program were. He had spent the day in his car out in the parking lot, which he was living out of, more or less, at the time, studying over the program. I asked him where he learned about computers. He said in high school and at MIT where he'd studied for two years before returning home to New Hampshire to help his mother on her farm. He said he'd given up the cold of New England in search of a little sun. I overlooked his generally scruffy appearance and asked him if he wanted a job. I always suspected that my newest employee moved out of his car to take up not only employment, but residence at Key Systems after hours until he had enough money together to rent a house trailer with a couple of the other programmers. But he did not disappoint me. Bob Davis proved—and continues to prove—to be one of the best hires I've made in all my years in business.

Though we had done some programming for manufacturers, including a job costing system and an inventory control package, we were not by any means solely a manufacturing software systems house. For a host of reasons, not the least of which was our location in the Keys, I felt we couldn't afford the luxury of being specialists. There was also the sense of pleasure we took in being mavericks for hire, willing to tackle anything and everything that would meet the payroll. We would do a job and move on to something new. The joy was in the challenge. Everything was new.

The sale of Software International, which I still had a financial stake in, to General Electric Corporation in 1980 netted me a considerable sum of money. I used this to acquire a building in order to move into more commodious quarters for our growing operation. About this time, I also instituted a more formal dress code at work, since cutoffs and flip-flops didn't exactly engender the level of confidence investors liked to have about where they were sinking their money. In Marathon, the employees of Key Systems began to radically stand out, being the only people to wear shirts and ties besides the local bankers and lawyers. It took some adjustment, but everyone seemed to accommodate the change well enough.

I invited Bob Davis into my office one afternoon for a performance review. Bob was doing excellent work for Key Systems. I overlooked the fact that on his lunch hour he would ride his bike down to the local nude beach for a midday swim, content with the fact that at work, at least, he had started wearing a tie. I said I wanted to give him a raise, but that I had an offer to make him. I was prepared to double the amount of his raise if he was willing to cut his hair. It wasn't mandatory, the choice was his. He didn't say anything at first. Then he said he'd think about it. I didn't know what he'd do. He kept me guessing for a couple of weeks, then showed up one morning with his ponytail shorn.

Life was changing, indeed, for us all.

The landscape of manufacturing in the American heartland was changing dramatically. Things set in motion over the previous twenty years were beginning to show obvious signs of distress in the American economy, greatly impacting the country's ability to compete in a host of traditional mainstay industries, from automobiles to steel, textiles, and electronics. Kennedy's wrestling with the balance of payments in the early 1960s, igniting a new round of foreign competition; the elimination of the fixed foreign exchange rate in the 1970s, weakening the dollar; and the simple fact that Asian and European manufacturers were attending to the details of product and process quality, production flexibility, and cost management much more diligently than American companies had radically tilted the playing field away from favoring U.S. production hegemony in the newly emerging global market. Inflation was up dramatically, making money for investing tight; productivity was down; and plant closings and corporate restructuring, coupled with outsourcing of production to foreign subsidiaries, was prevalent.

By the early 1980s, economists spoke of much of the American manufacturing landscape as being in "sunset," or lying in a "rust belt." In automobiles and consumer electronics, two mass-market industries we had invented, our position was severely weakened. Both were industry sectors that Asian and European manufacturers had long targeted, starting at the low end of the market where they could gain a toehold we weren't willing to defend, and thereby gaining not only a market beachhead, but an opportunity in which to practice and perfect mass manufacturing techniques.

The automotive industry was critical for its standing not only as the largest industry in the country, but a major customer to other industrial sectors, including alloy steel, aluminum, rubber, and machine tools. The number of imported cars was less than 1 percent in 1955, but would rise to become a third of all automobiles purchased in the 1980s.[30] The number of U.S. auto workers peaked at 21 million in 1979, and in 1980, the Japanese surpassed the Big Three in total units of production for the first time.

In 1955, nearly all radios sold in the U.S. were made here, but by 1975, the number was down to zero.[31] Sony introduced its first television, an eight-inch miniature monochrome set in 1960, competing against an industry dominated completely by American-made brand names. By the mid-1980s, the U.S. was down to one remaining manufacturer, Zenith, which held only 15 percent of the market.[32]

The upheaval in American production was fertile ground, spawning a small but rapidly expanding cottage industry of pundits, consultants, and gurus proselytizing various and sundry notions. Perhaps the one with the most cachet among Fortune 1000 companies at the time was MRP II. MRP II was the heir of Joe Orlicky's MRP—material requirements planning, which had come into prominence during the early 1970s. MRP II, which stood for *manufacturing resource planning*, had its roots in Orlicky's computerized time-phased material planning calculations in more than simply being a new play on an old acronym. MRP II had been coined in a meeting of individuals all versed with firsthand experience with MRP. The meeting was held in the living room of Oliver Wight's house and included Jim Burlingame, Walt Goddard, and others. Wight is credited with coining the term, preferring to leverage the established acceptance of the old term, yet distinguish the new set of production management principles and accompanying algorithms as being something different, new, more advanced. Wight and his organization were skilled and highly polished in promoting MRP II and, with the deteriorating state of competitive affairs in this country, had an audience among manufacturing executives quite receptive to hearing an uplifting message that had any semblance of promise and credibility.

Key Systems had grown to about a dozen employees by 1980. We had done enough work with manufacturers to keep our hand in the evolving MRP market. In addition to the job costing system we had done for a Massachusetts-based company, we had also done a bill of material system for an aircraft propeller manufacturer and one for Ithaca Intersystems, a small computer system manufacturer in Ithaca, New York. I enjoyed the work in the field in which I had started out, but didn't give it any special emphasis, feeling, I guess, that that chapter of my life was behind me. Yet serendipity in late 1981 would prove to the contrary.

One of my salesmen at the time, Dick Seibert, asked if he might exhibit at a local-area computer fair in Miami. It wasn't that expensive to obtain booth space, and although I didn't give the event much credence as a viable venue for new sales, the salesman was very enthusiastic about it. I authorized a bare minimum expenditure and told him to go ahead and go, but thought it was pretty much money down the hole.

I could not have been more wrong, yet more pleased, when it proved otherwise.

9 Breaking New Ground

One must necessarily discard information when one creates a concept.

Tor Norretranders, *The User Illusion*

Jesse Jones had a problem. Actually, he had two problems. Jones had been in business for himself since 1963, when he started a small job shop in the Miami area to manufacture service parts for aircraft. He had gotten into the machining of parts as a teenager to support his hobby racing cars and, as those things go, ended up with a shop of his own with a dozen or more machines and a couple dozen employees making precision parts for the aerospace, computer, and medical instruments industries by the early 1980s. Jones' first problem, one with which he had long wrestled, was how to efficiently schedule work for the shop. His second problem was that, at that time, there were no software packages on the market that began to address the complex set of problems unique to his environment.

To compound his problems, his largest customers—those in the burgeoning high tech sector—had all jumped aboard the MRP II bandwagon and were pushing jobs into his shop using their MRP II packages to dictate priorities, schedules, and costs. The CFOs and their staffs at these Fortune 1000 companies were keenly enamored with their system's ability to calculate inventory evaluations at the rise or fall of projected earnings. Their delivery priorities would subsequently be changed quickly, schedules pushed out or in depending on vacillating forecasts, standard lead times, and infinite scheduling—all of which served only to wreak havoc at Associated Machine Company, Inc., the small, lean job shop Jones ran in an industrial area on the outskirts of Miami.

Jones' fundamental problem in finding a solution was that he ran his shop not by forecast, but by orders. Though he did not call it such, he was what is universally known today as a make-to-order job shop. Nothing was scheduled until an order was placed—for there was no way of anticipating from one period to the next what product requirements would look like. The trick then and always in such an environment is how to effectively manage the capacity of the shop. You could always take orders for more capacity than you had, and you could always lose more customers than you wanted when you couldn't deliver as promised, but one thing you could never do was have more capacity than you actually had.

MRP II, driven by the dictates of Oliver Wight's industry-accepted prescribed standards at the time, did not embrace the job shop, make-to-order environment. The standard was make-to-stock. Make-to-stock, and subsequently the prescribed standard method of production management encoded into MRP II packages at the time, is driven by a forecast and/or master schedule. In order for MRP II to work efficiently—given the still prevailing limitations of computer technology at the time—the algorithms assumed infinite capacity. Infinite capacity was then—as always—an oxymoron. Nobody knew this better than Jesse Jones. And nobody knew better than Jones and legions of other job shop owners across the country that state-of-the-art MRP II did not address the requirements under which he was forced to operate. MRP II was virtually useless to him.

Being a precision-oriented professional, working in the realm of 1/1000 of an inch tolerances in the piece parts he jobbed for his clients, Jones was by necessity, if not predisposition, someone interested in the finite details of things. Jones had long known that what he needed to schedule and manage his shop was a system capable of working at a highly refined level of detail, especially regarding capacity. He had not been able to find such a computer-based system. Subsequently, he developed a manual system using multiple clipboards and a scheduling chart to represent his routings and the load on the factory floor. He called this his "Chinese computer system," borrowing from the notion of the Chinese abacus. Each clipboard represented a work center. As an order moved through the shop, its order sheet was moved from clipboard to clipboard, thereby keeping track of work-in-process.

At night, late into the evening after everyone else had gone home for the day, Jones would sit in front of his Chinese computer, studying the scheduling board and the status of work at each work center. An immensely patient man, Jones would sit and simply absorb the data that was before him, taking into his own dynamic memory the immediate permutation of work-in-process, then the orders outstanding on the scheduling board. Unable to find a software package to perform this task, he labored to approximate the talent of an idiot savant, someone "programmed," if you will, to excel at this very specific task. The challenge, he found, was not so much getting everything into dynamic memory, but keeping it there. Just about the time he was on the verge of envisioning the next day's schedule, invariably something broke his concentration—a phone call, a loud noise, an idle thought—and in one rapid instance, his dynamic memory of all the potential new permutations was scrubbed clean. Many a night, he left the shop with a less-than-desirable schedule for the following day's work.

Even with the best laid plans, running a shop is a highly dynamic endeavor. If the day's schedule at 7 A.M. was as near-perfect a permutation to maximize every resource to its fullest capacity, by 8 A.M. the reality of what had priority was invariably different. A customer would call to change an order; a tool would break; a machine would malfunction; material would prove deficient. Even if Jones had had an idiot savant to serve as the dynamic memory for his Chinese computer, he would have still been handicapped by the nature of his business. All his big Fortune 1000 customers were running their MRP II systems based on infinite capacity. Explanations and excuses to the contrary did nothing to promote confidence that he was a supplier they could count on.

Ever hopeful that there might be a computerized solution at hand somewhere, Jones paid the registration and attended the Miami-area technology fair. Though it didn't take Jones long to ascertain the salesman didn't understand what he was talking about, our salesman was smart enough to get him to agree to meet with me. My meeting with Jesse Jones proved to be fortuitous—not only for Jones and Associated Machine, but for Key Systems as well.

I liked Jesse immediately. He was intelligent, unpretentious, and knew what he wanted. I was also greatly curious about what he was telling me, how he had searched for a real, viable software solution to the problem of scheduling a job shop, where all work is associated with a customer order, not a forecast. He had done his due diligence and had talked with just about every MRP II software vendor he had run across, and although many of them professed they had the solution he needed, with a little further questioning he invariably determined that they did not.

Some vendors were beginning to understand that there was a huge untapped market in manufacturers who ran make-to-order operations; some had tweaked their make-to-stock-oriented systems to appear that they supported make-to-order, but ultimately they were simply the same horse with a different saddle.

Jones was rightfully intrigued by my curiosity and interest in his problem and in my background in manufacturing with IBM and Software International. Intuitively, I could appreciate what he was telling me, but I, like everyone else in the manufacturing software business, had been heavily swayed by the effective marketing of IBM, Wight, my own company, and others that there was a standard way to design and built an MRP system, and if you were going to be in that business, you did it the Wight way. Anything else was "outside the box," and suspect on the face of it.

Intrigued though he was, Jones was not convinced that I really could do what he needed to have done. That in itself was challenge enough for me to stay the course. I had always been spurred on by an intellectual challenge, particularly when someone didn't think I was capable. Jesse and I met and talked over a six-month period, me listening and asking questions, inviting him to go deeper into the workings of managing a make-to-order environment.

We talked all through the winter of 1981–82. Slowly, I began to truly grasp the significance of the difference between the two production environments, the one I'd taken my catechisms in and the one Jesse was revealing to me. Just as slowly, Jesse began to have real confidence that I could design and build a software solution that offered the best hope he had for managing the finite capacity of his resources. We both came to appreciate that in this endeavor, we needed each other. If he was going to have any hope of getting what he wanted, and if I was going to have any hope of providing it, it would have to be a collaborative effort.

Jones was savvy enough to understand that building software is an evolutionary process: you scope, design, and code the core functions, then add to them over time. He was willing, he said, to sign an agreement and front a significant sum of money to get the project going, if I was willing to ensure that Key Systems would never engineer a new release that didn't incorporate a viable migration path for him to elect to go forward. We signed the contract on April 13, 1982. Jones paid a one-time price of $59,700. I agreed to escrow the source code and personally agreed to never leave him behind on any future release I delivered.

Beyond the lawyers and the document with our signatures, we both had come to have a deep abiding respect and trust in one another's capabilities, and in each other's word. Jones was confident he was finally going to get the solution he needed; and I was back in the manufacturing software business with a zeal I had not known before.

What Jesse Jones wanted was a manufacturing system that had all the planning tools of MRP II, but one driven by orders, not a forecast. Jones needed to be able to estimate a job, take an order, schedule the shop, issue materials, record labor, ship, and post it all to a financial accounting system. It worked to the same summary, but started from a very different point than traditional MRP II.

Our contract specifically called for the ability to enter a shop order with customer requirements and due date, a bill of material structure of component requirements necessary to build a parent item, routing detail of all operation sequences and time requirements, and an inventory control function for managing materials from raw to finished goods.

Committed as I was, I still only partially understood the true nature of the adventure we had undertaken. I knew that what was required was a basic reordering of some traditional tasks outlined in Wight's standard model, starting with having order entry be the first trigger to the creation of a bill of material, which was subsequently chained to purchasing for acquisition of materials, followed with scheduling of the shop. I sent various programmers on my staff to meet with Jones after hours at his shop. Late into the night, he would walk them through the process of how he needed to manage things. Our challenge was that there was no real blueprint from a system's perspective for doing what he needed to have done.

We were chartered with building a totally new framework for manufacturing control systems (MCS), and we knew that it was wide-open territory for exploration. But we also knew it was a very large field we were attempting to encompass. We could do anything we wanted, in one sense, but we also had to be smart enough that we could capitalize on the broad market opportunity that lay before us—while making it all fit on an 8-bit computer. We started out writing code in BASIC, then shifted to Pascal, and ultimately wrote it in COBOL. The original hardware we started with was the TRS80, but subsequently we moved to the Altos 8000, which was more versatile and robust. It was unexplored territory, indeed.

At the heart of the problem were two issues: how to schedule work in a shop when there is no forecast from which to work and how to efficiently assess the impact of changing priorities on the overall load of the shop. Both of these concerns flew in the face of everything that I had ever done programming manufacturing software, not to mention everything that was widely touted as acceptable in manufacturing software design.

Wight's model for "closed loop" MRP II was, in fact, an evolutionary improvement over Orlicky's MRP design. It was considered a "closed loop" because the bill of material explosion within MRP was passed through the capacity requirements planning component to assess the feasibility of the master production schedule which had initiated the MRP bill of material explosion. In reality, however, the capacity requirements planning component was really not much different than what we'd devised years ago at IBM in blueprinting the PICS manual. Using production lead times determined by lot size rules within MRP, the capacity load for each work center was calculated by back scheduling from the requirements, due dates in MRP without any acknowledgement of any other work-in-process demands or previously scheduled requirements in that run. It was true that this so-called "closed loop" enabled the verification of conflicts in load given the existing plan, but the picture was inadequate from the start, assuming infinite available capacity.

Interestingly, Orlicky had seemingly intimated the problem with this type of algorithmic processing in concluding chapters of his seminal work on MRP, *Material Requirements Planning*, published in 1975. "The time has come to rethink certain traditional concepts, axioms, and theorems," he wrote. "Many of these are no longer relevant or valid because they fail to take into account the recent great enhancements in the ability to update for change. Traditional views that now must be revised pertain to the following topics: 1) manufacturing leadtimes, 2) safety stock, 3) queue analysis and queue control, 4) work in process, and 5) forecasting of independent demand."[33]

Standard leadtimes in manufacturing is an oxymoron and Orlicky understood that one of the critical problems with inaccurate lead times was the probability of the MRP calculation driving up work-in-process, impacting both costs and capacity load. "It seems clear that optimum leadtime values cannot be constant, as they are a function of capacity and the load pattern resulting from a particular product mix that is in production at any given time," he stated. He went on to argue that system design criteria for different business environments was a vital, but unexplored area of research. "Questions of planning-horizon length, time-bucket size, time phasing of allocated quantities, replanning frequency, etc." for varying production environments was an open question worthy of being "isolated and quantified."[34]

These questions, in direct contrast to what Wight had put forward in his software guide, were at the heart of what Key Systems was attempting to do in designing the MCS package for Jesse Jones. Fundamentally, one of the key issues for us—and one that flew in the face of what the entire software industry was then upholding as sacrosanct—was the relevance and critical importance of devising some method for recognizing the finite ability of a shop to execute a desired plan. Infinite capacity loading invariably overloads work scheduled through the shop, because it assumes the standard capacity of each work center without accommodating loads from previously processed orders, hence without considering any existing work already in process. Another way of thinking about finite scheduling is to view it as reality scheduling, for it takes into consideration the current load on the shop as it calculates a new schedule based on the input of new orders or work schedule priorities. What we were attempting, and what Jones desperately needed, was some way of dealing with the finite capacity of resources at any given time.

In accepting the contract with Jones, we were going against much of what Wight's organization, APICS, and other industry consultants were proclaiming as sound software engineering. APICS, pursuing a noble idea, was educating current and prospective members on the best practices for manufacturing and inventory control. In the 1960s, APICS's original reluctance to embrace outside consultants was based on their preference to maintain a separation of church and state, between the practice of production and inventory control and the promotion of marketing services to its members. There was merit to APICS's initial concern that led to their arm's length approach to consultants. The unmistakable success of the MRP Crusade of the 1970s, led by Oliver Wight and others, boosted membership and put the organization on more solid financial footing. Over time, the discipline and practice of production and inventory control became intertwined with a growing proliferation of interests, from the Big Eight, now the Big Six, to the smallest consulting practice. One certified the other, based largely on those principles endorsed by Wight, and then the other sold services back to the group, perpetuating the iron lock of Wight's early view of what warranted a standard MRP package and how it should be used.

We came to feel strongly at Key Systems, however, that we had no option but to go against what was sanctioned; anything short of this would be to concede that an entire domain of manufacturing, classified as make-to-order, lay outside the realm of viable software tools. It was not easy going. There was continual give and take, reexamination, and downright passionate disagreement and spirited debate among members of the team, and especially with Jesse Jones. Jones knew better than all of us what he needed, but he didn't understand the limitations of the hardware technology or software design principles that obstructed us from moving forward in a linear fashion. Difficult and trying though it was, the process of developing software for Jesse Jones was greatly stimulating and everyone involved in the project was keenly aware of that. We were breaking ground, laying track in an entirely new direction.

What Jones wanted was a solution to the conundrum of his Chinese computer: *What can I work on now that best meets my customers' schedules and keeps my shop running at peak efficiency? And when the reality of what is happening in the shop, coupled with new orders and new demands for pressing priorities, scuttles my plan, I want the tools I need to help recraft the most realistic, viable plan to keep the shop running efficiently and keep my commitments as best I can to all of my customers.*

No small undertaking.

In layman terms, what Jones was demanding was a system that would enable him to come to work each morning and reassess the lateness of the shop, and then take active steps to manage the work to the pressing criteria of each new day. "When I come in and find that what I am actually working on today is supposed to already be completed, I need a way to update the schedule so I can manage the ability to work on it today," he explained.

In his mind, the solution was to have the system give him the ability in the schedule to reconcile the growing shortage of days by adding days to the calendar of time *already passed.* It was untenable: there was no way, given the limitations of computing horsepower of the time, to rerun the schedule at the level of frequency he required—not to mention the impossibility of adding time to the past.

What we devised, and managed to get to work in a manner that adequately addressed the problem, required a certain shift in mindset to where the solution lay. The only solution was to add jobs somewhere further along in the schedule where there was room for them, given the capacity load on the shop. This at least gave you visibility to the actual load on the shop and permitted you to start work on what you needed to do, even if you were out of sequence on jobs scheduled later in the calendar. At minimum, you were maintaining the truth of the capacity that was actually available—and this was a definite improvement in the tools available to Jones at the time.

This was also the initial groundbreaking work at Key Systems in the development of a true finite scheduling solution. Our approach was significant for several reasons. The standard practice of infinite capacity loading ignored work that was already in process. It employed a backward scheduling algorithm, coupled with standard lead time offsets that enabled you to "pinpoint" a start date for the order, given the due date required by the customer. A compounding problem, however, was that the current work in the existing shop schedule was already behind schedule, meaning that you were scheduling in a phantom time slot, one that was already overloaded or expired—or both. This approach merely worsened the existing lateness of the schedule, such that the water was further muddied in understanding what the true nature of capacity of the shop was at that given time. As a result, work in process escalated and customer due dates became increasingly difficult to manage. The dates became irrelevant in running the shop. The fundamental objectives of the business—meeting customer order dates and managing performance at peak efficiency—were ever more distant and impossible to achieve.

Dave Layne did the initial work on the finite scheduling component of the system and got it to the point where we could validate our design. Because of our design, we were able to provide much more realistic dates to each job, and if nothing else, we were able to provide constant visibility to contentions for resources at each work center. This enabled Jones to understand the impact of changes to the schedule, make more reliable promises to his customers, and manage bottlenecks in his shop much more effectively.

Bob Davis subsequently ended up doing most of the programming of the scheduling system. He devised some critical innovations to improve the speed of the calculations and made the module run much faster, though it still warranted being run on only a nightly basis. Before Davis was through, he wrote at least three different versions of the software until he achieved the performance we deemed necessary to provide a viable solution.

We delivered a full turnkey solution—hardware, software, and networked peripherals—to Jones' job shop in Miami. Jones had become quite friendly with most of my programmers, especially Dave and Bob, and managed, in addition, to get them to engineer more than a few enhancements that weren't covered by the contract. When we got the whole thing up and running for him, he was quite pleased with the result of our collaborative effort. He was happy; we were happy. Key Systems had achieved the impermissible: we had developed the working foundation of a viable finite capacity MCS system tailored specifically to the requirements of make-to-order manufacturers.

The future looked exceedingly bright. And all of us at Key Systems felt we were only then just beginning to have fun.

10 "MRP—The Great Rip-Off"

Those anticipating distrust are more likely to undergird their conclusions with substantial statistics, or at least adorn them with false statistical finery.

John Allen Paulos, *Once upon a Number*

By 1983 the packaged application software market had grown to be nearly a 7.5 billion dollar industry.[35] Though much of that was in financial and human resource applications, many of the large independent mainframe software companies had begun to recognize the huge market potential for manufacturing applications. Much of it lay under their noses, existing as a "captured," untapped market already among their installed customer base who, having mastered financial accounting and human resource systems, were now clamoring for MRP II systems. Many of the large independent software suppliers had either recently jumped into the MRP II market through acquisitions or partnerships, or were looking to do so.

Though Oliver Wight died of throat cancer in the summer of 1983, the Wight organization would continue to promote the vision he propagated through the various educational and consulting forums it had perfected over the previous ten years. APICS membership had grown from its pre-MRP crusade level of 8,000 in 1970 to more than 50,000 members in 1983.

In its annual Man of the Year issue for 1983, *Time* magazine named the computer to the celebrity spotlight. IBM had decided to build a desktop, or personal computer, in 1980 and had announced that the design was complete in August of 1981. The IBM PC was delivered in 1982. It ran at the blazing speed of 4.77 megahertz and employed the MS-DOS operating system through an agreement IBM had signed with a fledgling company called Microsoft. IBM's foray into the desktop market was prompted by the huge success of Dan Bricklin's VisiCalc software package in the late 1970s, as well as the potential threat that Apple Computer posed to cornering the personal computer market with the Apple II. By the end of 1983, however, it was apparent that the entrance of IBM with its PC was what the market had been waiting for: it was an almost instant runaway success.

The early 1980s brought high tech to the fore and promised a wild and tumultuous ride for everyone in its path, from vendors to investors to venerable, staid industry service niches now threatened by upheaval to CEOs, CFOs, and corporate users across the spectrum. Lotus 1-2-3 was announced in 1982; the IBM XT with its hard disk drive followed in 1983. Apple shipped the Macintosh, the first mass-market personal computer, in 1984, the same year that the U.S. Justice Department broke up AT&T and General Motors acquired EDS for $2.5 billion to help it engineer its massive effort to computerize the manufacturing of automobiles.

Key Systems was riding a wave of its own. We did a million dollars in business in 1983, the year we delivered our job-shop MRP II system to Jesse Jones. In 1984, we decided to relocate from Marathon, Florida, where the cost of housing had become prohibitive for a growing company. We had grown to 23 employees—and were looking to hire new people all the time. We headed north again, to New England, but this time we settled in New Hampshire, where the dollar went a lot farther than in Massachusetts, due to the cost of housing and taxes. In 1985, our MRP II package, now known as the ProfitKey System, had gone through numerous minor and a couple of major enhancement releases. Due to its success and to confusion in the market for the fact that there were other business ventures operating under the Key Systems name, we decided to take the name of our flagship product as our corporate identify. Our first advertisement under the ProfitKey Systems corporate banner featured Jesse Jones in a full-page spread. The ProfitKey MRP II System, he declared, was the best machine tool he had ever invested in.

There was a sense by this time that we were a real company, a player, no longer just a cowboy outfit for hire. Of the 23 employees we had in Marathon when we announced our move, 20 elected to relocate to Salem, New Hampshire. Our business focus was now exclusively on manufacturing software applications. We began to organize into more discernible functional areas. We had a real customer service department, which John Houk headed. Our sales team was growing and expanding its reach. And we had clearly broken new ground in the applications that we had designed and built into the core product.

ProfitKey 500 brought the threshold of our system functionality to a new level of robustness. We had done some pioneering work with Anderson Tool, of Anderson, Indiana, to resolve one of the most pressing problems that had daunted developers—and haunted my memory of what I had helped originally engineer with the initial IBM PICS specification, back at the dawn of the computer software age. Anderson Tool was a make-to-order operation that manufactured large-scale equipment for other manufacturers, including, among other things, the machines used to make disposable diapers.

Though computerization of the bill of material inventory explosion process was an advance in efficiency in calculating material requirements, it had come with some very real and significant costs. Chief among these was the loss of a comprehensive audit trail for all inventory items, including where items came from, where they were used, their current status, and what jobs they had been assigned to or consumed by. Bob Sheldon, president and CEO of Anderson Tool, had walked me through his plant when we were first negotiating a software contract and showed me a huge warehouse stuffed floor to ceiling with material. None of it belonged to him; it all was material belonging to his customers, to the various jobs he had contracted to perform for them. But because he had no ability to assign, or peg it in his bill of material processing system, he had no way of effectively and efficiently documenting that it all belonged to customers' jobs.

In the eyes of the government, it was inventory on his books and, therefore, subject to taxation. For sake of convenience—and the ordeal of having to document it all—he typically assigned a value of $20,000 to $25,000 to the pile of material. If Bob Sheldon wanted one thing from a new manufacturing software package, he wanted to be able to assign and verify that this was customer job-specific material.

The problem was intriguing, penetrating to the depths of what I came to realize was one of the most insidiously fatal flaws of early software design. It had its origin in hardware constraints of the 1960s, when computer memory and speed were insufficient to handle the processing of a fully detailed bill of material (disk space was limited even for fully processing financial applications). In the 1960s, we did in fact design a "workable" solution we could sell. It is important to remember that the primary premise of the undertaking in 1963 was to sell hardware, and that the keys to the kingdom for closing sales were in the hands of the chief financial officers. The solution that was devised thrilled the corporate financial gatekeepers, for it addressed one of their chief concerns: the timely cost accounting of inventory. The design assumption based on the early hardware constraint, however, had never been revisited despite the ensuing significant advances made in the state of the art of computer memory, processing speed, and hardware cost reductions. That the devil to pay in the bargain lay deep in the bowels of the plant became apparent to me only when we began the system analysis phase for Anderson Tool.

It is difficult to appreciate the significance of something as seemingly innocuous sounding as "level-by-level" bill of material structures. By the 1980s, it was steeped thick in the catechism of accepted industry practices. It was in the bedrock of the APICS education and certification program, taught and tested, implemented and celebrated in countless "success stories" touting the benefits of MRP II. The concept and design of level-by-level bill of material structures, I realized, however, was based on a not-so-subtle ruse, a sleight-of-hand that dodged the real issue, one that traded appearances of a solution for the substance of one. It was shocking to me that it had not generated outrage and protest, that it had gone all but undetected for so long.

A bill of material is simply a list of materials required to make a product. Properly structured, the bill makes explicit the relationship of all the component parts required to construct a finished item. Manufacturing management has successfully exploited the concept of the bill of material with great results from the time of Eli Whitney and the advent of mass production at the beginning of the Industrial Revolution. The bill was originally an integral part of the "process sheet," the document that lay at the heart of all production, for it told all the workers not only what parts were needed (the bill), but also the resources, operation routing, and estimated time required.

When computers were first introduced in manufacturing in the 1950s, computer memory was so severely limited, prohibitively expensive, and difficult to program that the entire bill of material structure could not be recorded as a single data entity. The idea of computerizing the full scope of the "process sheet" lay in the realm of complete fantasy. The solution of the day was to break the bill into component parts, such that each part or subassembly became an individual bill and an individual shop order to be scheduled for production. As a result, the number of shop orders grew exponentially, as did the complexity of managing their production.

This was accepted, however, as a workable solution to what was then the intractable problem of limited computer memory. And when it was presented to the chief corporate financial officers during the sales cycle, they viewed it as, indeed, an excellent solution, for it gave them what they needed, which was a convenient way to maintain and calculate the cost of parts inventory. The impact on the shop, however, was profound.

Despite whatever the understanding the CEO and the chief financial officers had regarding the nature of the business they were in (making engines, appliances, pumps, or what have you), the shop was now relegated, in essence, to being in the business of making parts instead of a product. With the bill of material of a finished end item now broken up into multiple, or "level-by-level" bills of material, there was no easy way of maintaining the end-to-end relationship between all levels and the finished end item short of massive computer generated paper reports run frequently (a solution that was onerous on the shop management for the sheer volume of paper involved). A customer order for a diesel engine, for example, was sent to the shop as numerous shop orders for various component parts and subassemblies—all which were required to make the diesel engine, but without simple, explicit association, one part to another, or to a single customer order.

The lack of associated relationships between component parts to the final assembly greatly exacerbated the problem of shop scheduling. The only fathomable solution was to employ something termed "standard leadtime." Standard leadtime, an aggregate allotment for the manufacture of the complete product, was subsequently subdivided and assigned in factional allocations to each of the various parts. But as everyone who works in production knows, standard leadtime is an oxymoron, so using it to schedule the shop is a faulty method on the face of it. What was worse, however, was that the assignment of the factional allocations to subassembly parts was wholly arbitrary, typically left to data processing personnel responsible for assigning values to the data fields in the part master file, a file that served financial accounting requirements far more than production.

The problem of when to start work on each item was still left unresolved. To address this issue, software designers added the ability to schedule the shop by backward scheduling each component assembly from the customer due date, applying the standard leadtime offset at each level of the bill. This only further compounded the deleterious effect of the whole complicated machination for two reasons. First, back scheduling totally overlooked any current load, or work in process, in the shop; and second, inevitable slippage at one level got successively compounded through all the subsequent levels. Though it all could be made to look good on paper—and paper was something the shop got daily by the reams in computer-generated shop reports—the net effect was that there was no way that anything was ever going to get made inside the standard lead time. It was, in essence, a cure that typically killed the patient.

Paradoxically, despite the labyrinthine nature of this so-called solution, it all made perfect sense to the financial officers. During the sales cycle when they were being wooed by adept presales system consultants representing the system vendors, the financial officers were typically curious about only one thing: Will this procedure provide me with the ability to calculate inventory costs? The answer was always an unqualified **yes.**

The disregard and open disdain that American corporate officers had for manufacturing production personnel in the 1950s, 60s, 70s, and even well into the 1980s was so pervasive and prevalent that there was little impetus to seriously consider the impact of these system machinations on the operations of the plant floor. Shop managers were portrayed in sales training programs at Fortune 500 companies as bumbling, near-imbeciles who couldn't find their way through a computer printout if they were led by the hand. It is little wonder that system designers were empowered to disregard the gravity of the impact of these "solutions" on the operations of the plant.

That much of these design liberties were canonized by the production trade associations and the manufacturing management gurus of the day is almost beyond comprehension. The cost—not to mention the disservice heaped upon companies struggling to be competitive in an increasingly globally competitive world—is truly staggering to consider.

And I was not without fault. I, like everyone else in the industry, had accepted many of the assumptions behind the rationale of the initial system design, going back to my involvement in the original PICS design team at IBM in the 1960s. But beginning with the work we did for Jesse Jones in the early 1980s developing an MRP II system specifically designed for the MTO environment, and in the system design initiated by our work with Anderson Tool, I felt I was involved in—and committed to—a significant effort to remedy some of the most egregious production management fallacies perpetuated in the rush to sell computers to manufacturers.

When we delivered ProfitKey release 500 to Bob Sheldon at Anderson Tool in 1984, we provided him with a multipegging bill of material inventory system, such that he could document every piece of material from finished goods down through all of the multiple levels of the bill to the last bolt and screw belonging to a specific customer contract. This, coupled with our increasingly robust finite scheduling capability, made the ProfitKey System stand without parallel in the entire array of competing MRP II applications. In addition to the job shop market, we now had a system that addressed the complexities of custom manufacturers, those who made finished products of variable and diverse, and typically deeply indented, multiple-level bills. We had a winner, and in the middle 1980s, we were riding it hard.

Which is not to say there weren't problems in paradise. There were plenty. Some were beginning to make themselves felt in the manufacturing software market, which played to our favor in the near term; and some in the less obvious, underlying core technology behind what the user saw on the screen, that we were—if not oblivious—woefully myopic in picking up on. Time would show that we were well positioned to take advantage of the first thorn-bed of problems striking the industry, but had plenty of company in suffering the fate of the second—not that shared company would prove to ease the pain for me, personally.

By the middle of the 1980s, something of a backlash was building in the market against MRP II. Unsubstantiated claims in 1985 by a consulting firm that the "success rate of MRP isn't good, it's astronomical"[36] prove curious in light of an article in the November/December 1985 issue of the *Harvard Business Review.* "International Data Systems estimates only 25 percent of material requirements planning systems have achieved installation objectives," the article stated. More specifically, the article went on to say that the "techniques for inventory control have become so esoteric and complex that some are either unusable or can be easily misused." (Interestingly, singled out for particular note was a highly problematic exponential smoothing formula error that originated in 1965, one that had been replicated in various publications and journals as authentic in up to 23 instances. The application of the term "exponential smoothing," you may remember, was my grand contribution to the obfuscation required in the original PICS manual to smooth over—pun intended—the severe hardware speed and memory limitations of the time.[37])

Though companies were spending in excess of a third of their total durable equipment budgets on information technology in the mid 1980s[38]—much of this focused on increasing manufacturing competitiveness—there was a lot of grumbling and complaining. The explanation for the difficulty of achieving implementation objectives and blame for what was wrong with MRP II erupted into vigorous claim and counter-claim. The reasons (or excuses, depending on one's perspective) were legion, ranging from management issues to planning to analysis and systemic causes.

The July 3, 1989 issue of *Industry Week* magazine looked at the problem and quoted Ken Stork, computer director of materials and purchasing at Motorola, Inc., as saying, "MRP II became a crusade pitched by a welter of vested interests.... It took on an aura of importance well beyond its ability to crunch numbers...." In counterpoint, Walt Goddard, who was then president of Oliver Wight Companies, added an assessment that I would subsequently champion wholeheartedly. Goddard stated, "The CEO thinks MRP

II applies to purchasing and manufacturing, but typically doesn't see how it applies to him." Though our assessment might be the same, my reason for arguing this would probably vary greatly from that which prompted Goddard to make the statement at the time.

I would hear a startling claim at an IBM partners meeting I attended in 1989 that spoke volumes to me of the bottom-line impact of much of the pain and recriminations heaped upon MRP II. I don't remember the gentleman's name, but he was a big, burly man with red hair. He said that 87 percent of all MRP II projects are terminated within six months. More distressing still, he stated that 40 percent of the Fortune 500, as listed in 1979—ten years earlier—were not only not on the list, but no longer existed at the time of the 1989 accounting. This was a staggering revelation to me, and one which I'm convinced spoke to the heart of the backlash that had been raging over MRP II.

I had become by that time an adamant believer that one of the fundamental problems with MRP II, as it was generally ascribed to, had to do with the issue of capacity planning. A September/October 1985 issue of the *Harvard Business Review* agreed with my assessment. The article claimed that there were somewhere between 2,000 and 5,000 companies using MRP at that time. The author went on to say that "MRP… assumes unlimited capacity in all work centers, whereas in reality some work centers always behave as bottlenecks. This contradiction destroys the accuracy of MRP scheduling logic and makes it ineffective for capacity planning and control." The resulting consequence of escalating work-in-process and out-of-control inventory levels, I believed, would be found as systemic problems at the vast majority of that missing 40 percent of the Fortune 500 listing in 1989.

An adamant believer in this analysis—and a true believer in the value of finite capacity planning, with case examples in my customer base to back it up—I took on something of a crusade of my own in the last half of the 1980s. I was, it is true, something of a wolf—a lone wolf at that—howling in the wilderness, but it was a howl that resonated soundly among manufacturers at the time.

In 1988, I had our advertising people develop an ad for ProfitKey that won me no plaudits from my competitors. Never one known for excessive tact, I was blunt in my affirmation. The banner headline read: "MRP II—the Great Rip-off."

I created the notion of Infinity Airlines in the 1980s to try to explain the stupidity of infinite scheduling. In my seminars, I would explain that infinite scheduling could cause double and triple bookings of airplane seats; with the accompanying double and triple delivery of meals to the unsuspecting airplane and crew. It was such a hit that Frank Norris, my West Coast salesman at the time, had a commercial artist and PR genius named Don Stone draw this cartoon.

It caught the eye of the entire manufacturing industry. It generated immeasurable ire among my competitors, who countered primarily with the charge that it was only the latest proof I was dealing with less than a full deck. But the eye of the storm that blew among software vendors was nothing compared to the response I got from manufacturers faced with making product deadlines. The response was phenomenal—and almost all positive.

I had another ad developed that was based on an idea of mine. The ad included a cartoon that featured a fictitious airline, which I called "Infinity Airlines." The cartoon pictured what happens when an airline company looks to book its capacity of a 300-seat aircraft, but never acknowledges reservations as they are recorded over time. All is well with each passing day as travelers call to book a seat and the airline answers, that yes, there are 300 seats. But come departure time, when 600 people show up, the problem of infinite capacity becomes abundantly clear. It doesn't work.

This ad ran in the August 21, 1989, issue of *Industry Week* magazine. John Sheridan had an article in that issue of the magazine entitled "Preaching the

Gospel of Finite Scheduling" that featured an interview with me. The opening ran as follows:

"Richard T. (Dick) Lilly has this piece of advice for U.S. manufacturers: 'Don't lie to your customers.'"

Sheridan went on to quote me saying: "This country has a balance of payment problem for one reason.... It's because we can't deliver a quality product on time because we don't **plan** to.

"We promise to. We lie and say we will... But we don't **plan** to."

I did not fail to recognize the irony that this was an issue I had been party to in the debate around a conference table in White Plains in 1963. But time, as they say, brings wisdom. For some. Finite capacity scheduling was anything but widely upheld as a viable algorithm in 1989, even though my company had been delivering the functionality since 1983.

Capacity Requirements Planning, as it had long been taught by APICS and leading industry consultants and educators, put the ultimate responsibility of resolving conflicts between the master production schedule and shop floor capacity in the hands of the planner. The master schedule was blown through material requirements planning, which included processing of a data field in the item master representing infinite work center capacity. As it was a rough picture of things, the industry at least had the prescience to start calling it that—*Rough Cut* Capacity Planning—somewhere after the original inception of the concept.

Finite capacity planning had been around for some time as a concept—and since 1983 as an actuality, with Jesse Jones taking delivery of our software. But it was never embraced by the software industry and leading consultants as anything more than a nice idea—in theory. In practicality, it was deemed next to useless for the impossibility of building a scheduling algorithm that could automatically generate the calculation in a timely manner.

In truth, it was viewed as competitive to MRP II, and was subsequently deemed as lying outside the sanction of what was acceptable. Even though ProfitKey had proven that it was not only feasible, but possible to manage the resolution of capacity scheduling conflicts by a computer-based calculation, controversy and skepticism would remain attached to the idea in some fashion or another for some time to come.

As nature abhors a vacuum, industry consultants, pundits, and gurus were not without other nostrums and elixirs to promote throughout the 1980s. Just-In-Time (JIT) inventory practices, imported from the Japanese in the early 1980s, were touted as the solution to the deficiencies of MRP. Where MRP II was weak on the shop floor, JIT held the promise and would prove to be a great tool for managing execution of orders in the plant. At heart,

JIT called on simplification of design, processes, and routings in advance of—or certainly concurrent with—its implementation. Where it was cast first as a competitor and a threat to MRP II, it was ultimately subsumed as a critical complement to it.

JIT was followed by "computer-integrated manufacturing" (CIM) as the Holy Grail of manufacturing in the mid-1980s. CIM also had a bottom-up orientation, starting with the vast investment manufacturers already had in programmable logic controllers (PLCs) that directed machinery on the shop floor. The idea was to wire all this technology-driven hardware into one giant computer system, in order to build "the factory of the future," which was viewed futuristically as a "lights out" environment, where customer orders were poured in the front end and finished products dropped off the conveyor line at the shipping dock.

Despite mammoth investments by General Motors and the efforts of its EDS subsidiary and others, this technological Gordian knot proved too difficult to unravel, and CIM was subsumed by "World-Class Manufacturing" toward the end of the 1980s. The definitions behind each of these successive waves of high-flying banners seemed to get less specific—and grew more distant from technology—as time went along. Perhaps this was due to an increasing caution on the part of consultants and pundits to throw their hats too neatly into one ring. The National Center for Manufacturing Sciences defined a world-class manufacturer in 1989, as "simply ... being better than almost every other company in your industry with at least one outstanding aspect which gives the company a competitive advantage in the market place."[39]

Vague as things became, however, MRP II managed never to be too far behind the first rank of advancing proselytizers. "Just In Time in the MRP II Environment,"[40] "CIM Starts with MRP II,"[41] and "MRP: the First Step Toward World-Class Manufacturing"[42] are indicative of a whole raft of articles published in industry trade journals during the 1980s.

Upheaval and confusion in the marketplace? For sure. The Manufacturing Futures Survey, an annual survey conducted by the Boston University School of Management, traces among other indices the top five strategic programs for each year. In 1984, production and inventory control was top of the list. By 1986, it was down to number five, and in 1988, it was gone. In the 1988–89 time period, integrated manufacturing systems were listed in the middle of the list of *least effective* programs in the eyes of survey respondents.

Even though things sounded entirely bleak, there was growing room for hope in American manufacturing by the close of the 1980s. Ford and Xerox in the two bellwether industries, automobiles and electronics, had undergone huge transformations engineered by tough management decisions, careful cost management, and successful manufacturing design and production improvement initiatives. Ford rebounded from a $1.5 billion loss in 1980 to a $4.6 billion profit in 1987. Xerox cut its manufacturing costs in half and made dramatic product quality improvements over roughly the same time, netting a reversal in decline and enhanced market share. Xerox had invented the modern copier, but by 1979, its production costs and product development time were twice that of its Japanese rivals. But taking a lesson from its competitors, it simplified both product and processes, and marched back to be a competitor on par with the Japanese by the end of the decade.[43]

Things were bright enough by the start of 1987 that *Business Week* magazine ran a cover story entitled "Why Manufacturing Will Revive." The article's lead was teasingly cautious, however. "Don't hold your breath," it began. "Don't bet the ranch. Don't uncork the champagne just yet. But it does look as though maybe, just maybe, the nation's long suffering basic manufacturing sector is turning around."

11 The Albatross of Shifting Technology

Computers are dumb and can only do a few things. But they make up for it with speed.

George Gilder, *Microcosm: The Quantum Revolution in Economics and Technology*

Things were looking up for American manufacturing as the decade closed. The tumult in the technology sector, however, was anything but over. Disparate events that had occurred over the previous twenty years were beginning to coalesce into a critical mass that was on the verge of producing sweeping change in the technology landscape. Some players were posed to reap gigantic gains. Others seemed destined to suffer the fallout from the tornado generated by what might best be characterized by a marketing slogan that would grow to increasing prominence through the 1990s. The coming revolution of the 90s could be summed in three words: "*better, faster, cheaper.*"

The equation of change was working from both ends toward the middle, from both the hardware and software sides, though advances in hardware were more accelerated and obvious in their impact—at least initially. The changes on the software side came more slowly, like the drift of tectonic plates, but when the fault lines shifted, the upheaval was equally seismic.

The history of the transformative process of hardware technology development is compellingly chronicled in George Gilder's *Microcosm: The Quantum Revolution in Economics and Technology.* Gilder, an accomplished writer on economics and a Harvard Fellow, published *Microcosm* to generally rave reviews for its breadth of grasp of the sweep of the technological revolution of the twentieth century—the book appearing in 1989, just as things approached the cusp of commercial realization of "better, faster, cheaper." I take liberty in citing from *Microcosm,* as Gilder has not only the grasp, but the sense of the dramatic in his depiction.

Early in the book, in a chapter entitled "The Prophet," Gilder recounts the early research efforts of Carver Mead at the California Institute of Technology in the late 1950s. Extending work on the Nobel Prize–winning research of Leo Esaki on the tunnel diode, Carver's initial effort to perfect the tunnel diode ended in almost abject failure. "But his failure bore fruit more important than any prize. Still in his early twenties, Mead found in this flawed device the secrets of the quantum era and led the way into it," Gilder writes:

> Named from the Greek words meaning two roads, an ordinary diode is one of the simplest and most useful of tools. It is a tiny block of silicon made positive on one side and negative on the other. At each end it has a terminal or electrode (route for electrons). In the middle of the silicon block, the positive side meets the negative side in an electronically complex zone called a positive–negative, or p-n, junction.[44]

Being charged as such, it serves to conduct current from positive toward negative on one side, and the reverse on the other (hence, two roads). Scientists had long known, however, that if you apply a strong enough voltage against the grain, the junction would burst under an "avalanche breakdown" and effectively switch the flow polarity, confounding basic laws of electricity.[45] "Magic or not, however, the tunnel diode was both scientifically intriguing and commercially exciting ... the Esaki burst effect promised extremely fast switches, approaching the speed of light. All things being equal, the faster the switch the better the computer, which uses vast arrays of on–off switches to perform its high-speed calculations."[46]

Commercialization of breakthroughs in computing technology have roughly followed a twelve-year maturation cycle. (This is not to be confused with Moore's Law, which was popularized in the 1990s, which holds that computing horsepower doubles roughly every 18 months.) The twelve-year cycle represented the time it took from scientific discovery to true commercialization. The transistor, the integrated circuit, and the microchip all seemed to follow this principle, more or less. Also, the tunnel diode.

Mead, Gilder writes, pursued perfecting the tunnel diode "for close to a decade," attempting to build faster diodes, all without much success. But "Mead gained an intuitive sense of the quantum domain and reached an amazing conclusion for all semiconductor electronics. The industrial world might be telling him to invent a faster diode. But the technology was telling him a way to transform the industrial world."[47]

Throughout the 1960s, research centers around the country, "from Bell and RCA in New Jersey to IBM in Yorktown Heights and the research centers of Silicon Valley,"[48] were encountering continual resistance to breakthroughs in accelerating the speed of the tunnel. They all had to reach, they surmised, the impenetrable barrier in narrowing the tunnel at roughly one micron, or one one-hundredth the diameter of a human hair. The thinking held that any smaller than that, the sheer heat of further miniaturization would melt the circuit.

Mead, now deep in the microcosm (below one micron), came to the rather startling conclusion that "everything gets better as it gets smaller, cooler as it gets faster, cheaper as it gets more valuable."[49] This discovery, known as Mead's law, moved not only technology, but also economics and industry into what Gilder proclaims the age of the microcosm.

Gilder goes on to write that "the fall of the macrocosmic computer did not turn into a rout until that amazing day in 1971 when Intel announced not only the DRAM for working memory and the EPROM for software storage, but also a microprocessor, absorbing the entire central processing unit of a computer on one chip." This development, known as "Intel's triple play," was the sea of change necessary to fully launch the technological revolution of the 20th century.[50]

This breakthrough would roughly follow the twelve-year cycle before it was truly commercialized with IBM's delivery of its PC in 1983. Though we at ProfitKey Systems were keeping more or less cognizant of these advances on the hardware side and taking advantage of them as best we could, we failed, like many other software vendors who had been laboring long in our discipline, to fully appreciate some of the ensuing changes that these hardware developments were fostering in our own domain.

Microsoft was a winning success story in the late 1980s, with its DOS program resident on the vast majority of PCs sold around the world. But it was nothing like what it would become, despite its release in 1985 of its next-generation operating system, known as Windows.

The relational database had made its commercial appearance in the early 1980s in the mainframe computer environment, replacing the hierarchical, or "tree-like" database design with the more flexible relational model for "chaining" relevant and related tables of various data sets. Relational technology was augmented with a new structured query language (SQL) that was much more user-friendly, further boosting its appeal. Additionally, fast-rising competitors were taking advantage of developing their systems in what were deemed fourth-generation languages (4GLs), which were also easier to work with and thereby more productive.

ProfitKey remained exclusively a UNIX-based system, exploiting the much greater robustness of UNIX over the MS-DOS operating system. UNIX provided greater interoperability between systems and devices, as well as much greater scalability and reliability in large multi-user, mission-critical business applications. We had our own proprietary hierarchical database, designed for the maximum efficient handling of our hugely data-dependent manufacturing application. We were also still writing our programs in COBOL.

Our eye remained tightly focused through the late 1980s on continuing to add broader and deeper application functionality to our product in order to remain unparalleled in the business value we offered prospects and customers. As time would prove, with the sea of change in technology sweeping underfoot, we myopically focused on this value proposition to our detriment in the longer term.

The rapid rate in change of technology poses a constant threat to vendors and users alike, with the very real question of how to keep abreast of it all without spending a fortune in human and capital resources. At the same time, it also creates tremendous opportunity for new entrants to the market who can capitalize on advances without the invested cost of mature technology and an established customer base. The dark side of this scenario was what ProfitKey woke up to at the close of the 1980s.

We were finding small, entrepreneurial start-ups—like we had once been—competing against us in deals we thought they had no right to be party to. There was no question that ProfitKey had the richer, broader, deeper application set. And yet we were repeatedly asked about fourth generation languages and relational databases. From our perspective, fourth generation languages were yet without industry standards and, therefore, problematic as a reliable tool. And why would you want a slow, inefficient relational database when our database is fine-tuned to maximize performance of our applications?

By 1990, I was keenly aware that we needed to do something to address these issues. The embodiment of "better, faster, cheaper" in the competition we were increasingly encountering—and losing to—had created a situation that was clearly untenable.

The board of directors at ProfitKey, which included representatives from our venture funding sources, were becoming increasingly impatient with the slow deterioration in profits and market share. They were subsequently becoming more impatient with me and far less willing to hear me out on what I believed we needed to do to counter the situation.

The only way out of this mess was through it, I told them. We needed to fund and invest in developing a next-generation system that took advantage of the reigning technology advances, such as Windows and relational technology, while preserving our deep manufacturing domain expertise resident in the current ProfitKey application.

The board increasingly saw otherwise. What they wanted was better return on investment—now. The deterioration of the relationship between them and me paralleled the further decline in profits as the development strategy stalemate continued.

The board elected to bring in a new CEO in 1991. The new CEO had the mandate of the venture capitalists on the board who wanted to see increasing return on investment in the short term.

Things went quickly from bad to worse over the next several months. I knew what was needed, but was effectively made ineffectual in guiding the company I had started and built with the assistance of some very creative talent.

I despaired.

And then on May 1, 1992, I was summarily let go. I suspected it was coming, so it wasn't a total surprise. But I was devastated. The company I had poured my heart and money into for the last twelve years, the company that had pioneered breakthrough innovations in finite scheduling, MRP II for the make-to-order environment, and multilevel bills of material was no longer home to me. I was let go without any form of severance.

I was down, but I didn't consider myself out. Not yet anyway.

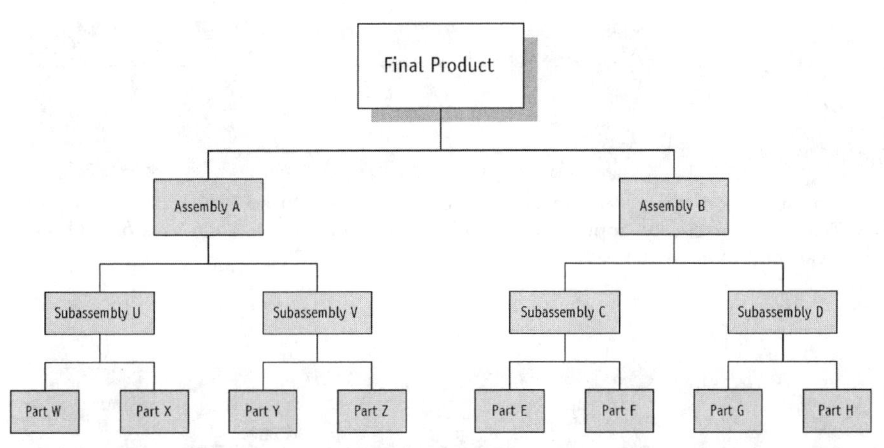

This is an example of a bill of material for a finished product. In this example there is a 1 to 1 relationship of parent and child. This means that to make one Subassembly U, one Part W is needed and one Part X is needed.

In order to manufacture a Final Product, both an Assembly A and an Assembly B must together be worked on at Resource 1, then at Resource 2. Each of these two operations can occur at a rate of 62.5 units per hour. To make 1000 Final Products from As and Bs would require 16 hours of Resource 1 capacity and then 16 hours of Resource 2 capacity.

To make 1000 of Assembly A and 1000 of Assembly B from their components takes 16 hours for each set of assemblies at Resource 3, then 16 hours at Resource 2.

To make 1000 of Subassemblies C, D, U, and V from their components takes 16 hours for each set of subassemblies at Resource 4, then 16 hours at Resource 2.

In a traditional MRP or ERP system, the Final Product, each assembly (A and B), and each subassembly (C, D, U and V) would be manufactured on separate work orders, each with a different due date.

In this example, "today" is 04/17/2001. Imagine that you wanted to complete 1000 Final Products by the end of the day on 5/11/2001.

In this example, a "standard lead-time" of one week has been established at each level. This means that it takes 1 week to receive the raw materials (E, F, G, H, W, X, Y and Z) after having been ordered. This means that if the materials are ordered today, those materials should arrive some time on 04/24/2001.

The manufactured parts (Final Product, A, B, C, D, U and V) have a standard lead-time of one week. This means that if we want to have 1000 Final Products by 5/11, we will need to have all the assemblies needed (Assembly A and B) completed by 5/4, which means we need to have the Subassemblies completed by 4/27.

The illustrations on the next page show typical scheduling scenarios, both with infinite and finite scheduling.

This first illustration shows the capacity of four resources with no load for the specified days. The units of capacity appear as white blocks; these resources each have one 8-hour unit of capacity, Monday through Friday.

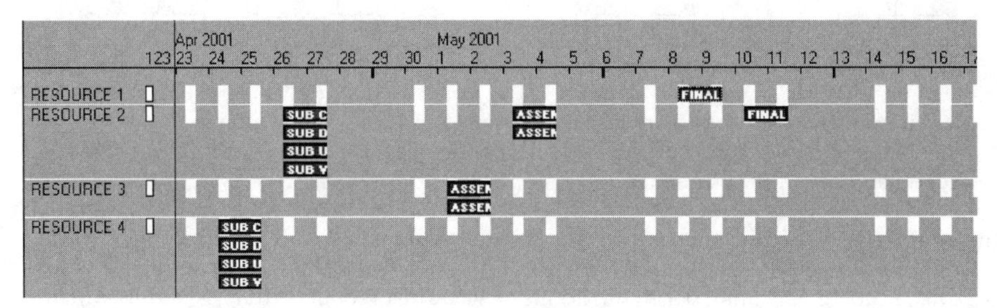

Traditional MRP, with infinite backward loading, attempts to schedule the last operation as late as possible in order to meet the want date. But traditional MRP does not recognize that capacity is not unlimited at the various resources; therefore, you can see that the system has scheduled numerous operations at individual resources, all to begin at the same time. Based on the lead-times, Final Product is loaded to complete by 5/11, Assemblies A and B are loaded to complete by 5/4, Subassemblies C, D, U, and V are all loaded to complete at Resource 2 on 4/27 and Resource 4 on 4/25. But the resources do not have the capacity to handle all of these operations at once. MRP leaves the user to handle resource overloads and to manually create an achievable schedule.

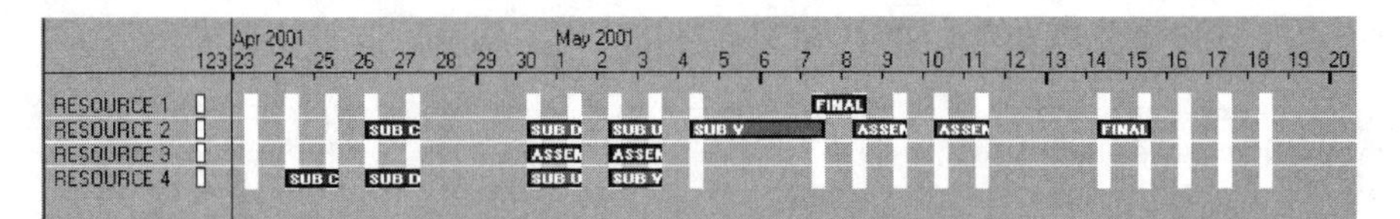

In this example, the work orders are scheduled with finite capacity, so there is no overloading of resources. Notice that Subassemblies C, D, U, and V are now scheduled sequentially at Resources 2 and 4. Notice however, that in this example the Scheduler does not recognize material requirement dependencies between work orders. Because each Assembly, Subassembly, and the Final Product are being planned to be made on separate work orders, each work order has its own independent completion date. With finite scheduling, the Scheduler is trying to achieve the independent targeted completion dates for each work order and is unaware of the dependencies between these work orders.

This illustration shows the same resources and schedule, but from a weekly view. You can clearly see that Final Product is scheduled to start before the Subassemblies that feed it are going to be completed.

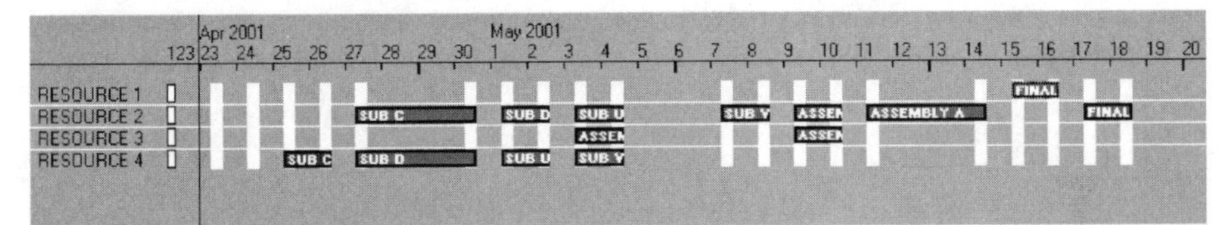

In this final example, the entire set of work orders is scheduled with finite capacity and material checking. Notice that there is no longer contention at the Resources (as in the prior example) and that each operation begins only after the supplying operations are scheduled to be completed (this includes supplying operations from other work orders, as well). This schedule shows that Final Product will finish one week later than originally predicted by the traditional MRP schedule (5/18, as opposed to 5/11).

The illustration below shows the same resources and schedule, but from a weekly view. Final Product is scheduled to start after the completion of ALL of the Subassemblies.

SECTION III:
COMMON SENSE MANUFACTURING

12 The Road to VISUAL Manufacturing

A team of two, with one leader, is often the best use of minds.

Frederick P. Brooks, Jr., *The Mythical Man-Month*

In addition to the radical changes in computer technology wrought during the closing days of the 1980s, or perhaps because of them, industry watchers began casting about for a new framework upon which to hang the collective impact these changes were having on the whole field of software development and deployment. In the early 1990s, Gartner Group began referring to the whole manufacturing software market by a new moniker: enterprise resource planning, or ERP. Other groups began calling it by various other names, as well, and something of a "branding" battle ensued for several years between various factions. In fact, there was little substantive disagreement on what was involved, beyond nuance and definition of terms. Despite the fuss, ERP was basically MRP II—only "more so." In addition to more robust links with other enterprise systems, from financial accounting and human resources to distribution, ERP was distinguished by global deployment requirements, including multi-language, multi-currency, and multi-plant (what came to be referred to simply in marketing parlance as "multi-multi-multi") functionality. Additionally, ERP was characterized by the new technology on which it was to be built, including client/server architecture, graphical user interfaces, SQL relational database technology, and 4GL-development.

Many software vendors made the shift rapidly—in name only. ERP had the distinct benefit from a marketing perspective of minimizing the continued association with MRP II. MRP II was quickly shunted aside in a flurry of printing new brochures carrying the ERP imprimatur. True, some went to the trouble of adding GUI "front-ends," which, in essence, simply put a pretty face to the same old character-based, "green-screened" systems. And a few companies began a more thorough reengineering of their systems. But in my mind, the real opportunity lay in starting from a clean slate to redesign from the ground up a truly new class of manufacturing software. This was the vision that inspired me to make that long drive to Ocean City, Maryland, Memorial Day weekend of 1992.

Without a job, without a safety net, time was a luxury I could not afford. Yet it was imperative in my mind to make the trip to Ocean City to personally meet with Buddy Damiano, Dave Layne's father. Dave saw it as a mere formality. He was confident his father would support him during the development phase of executing our vision for a manufacturing system that leveraged our expertise *and* took advantage of the new PC-based technology. I, however, felt it was critical to meet with him face to face; I was asking him to show faith not merely in his son's talents, but in my vision as well.

There very definitely was an element of risk involved. We had, at best, twelve months to design, build, and close our first sale before we either ran out of money or lost the market opportunity—or both. But if we could do it, and I wholeheartedly believed we could, the potential payback was enormous. I wanted to sit down with Buddy Damiano, have him look me in the eye, and see that I was absolutely committed to making this venture a success.

Buddy Damiano's only issue was to affirm his son's best interests. He told me he had no reason to suspect anything other than the best intentions, for he knew that Dave trusted me, that he suspected Dave even looked upon me as he would a father. There was truth to this, I suppose, and where I considered it a compliment of the highest order, I was even more impressed that David's father felt no threat in this. He told me he would gladly cover his son's family expenses. I encouraged him then to take the opportunity to invest in the venture for himself as well, for I was confident that Dave and I together could pull this off. It was deeply satisfying and affirming when he didn't hesitate to take me up on the offer.

David stayed with his parents through the rest of the holiday weekend, but after concluding the purpose for my trip, I began the long drive back to New Hampshire, my mind filled with an endless stream of ideas and issues that needed immediate attention. A year to develop a viable product and make our first sale was an aggressive schedule, and there was no luxury to miss a single day of effort toward achieving our goal.

Our goal in launching Lilly Software Associates, our new company, was to be the first software vendor to come to market with a solution based on a totally new development paradigm. Our focus was clear: to develop a Windows-based manufacturing software solution for the job shop, make-to-order market segment. This was the segment of the industry we knew the best—arguably, better than anyone else because of the years of pioneering efforts we had poured into previous developments. In addition, we wanted our solution to go beyond the intuitive characteristics inherent in the Windows environment. We wanted it to be "natively" intuitive to the manufacturing user in mirroring how manufacturers actually envision and perform their jobs.

I wanted to create a wholly "visual" manufacturing software system, one that people familiar with running a manufacturing operation, from shop workers up to the CEO—and *especially* the CEO—would feel comfortable with immediately upon sitting down in front of a computer to perform their jobs.

And while our vision was clear, time was still the critical factor. Despite our best efforts to be productive and proficient, I knew we were going to rapidly burn through the small amount of seed money we had at our disposal. The 12-month window I had given us was really the outside parameter for success; I knew we needed to be looking for our first customer to sign a contract in as little as five to six months. To be successful, we needed to design a strategy and execute faultlessly where the sales cycle was as short as 30 days. We simply didn't have time to go through a long courting ritual with customers if we were going to generate a sustaining revenue stream as quickly as we needed.

Due to my experience in the business, I had adopted three cardinal rules to successfully sell into the job shop market. First, for speed in closing the deal, the CEO needed to be the one who made the decision, for invariably, the more hands involved, the longer it took to close. Second, the ease of use and utility of the software needed to be immediately apparent from the first moment I sat down to demo the software to a CEO. It had to model the way operations people think and make decisions, and it had to be easy to navigate to check and validate all the permutations of inquiry required to make timely decisions. And third, and perhaps most significant, the price point needed to be deftly set. In my mind, this required making the value of the software so obvious the CEO could assess the merit on the fundamentals of improving his business. Given the option, would he buy the software for the shop or a new Mercedes Benz for himself?

It had to be based on client/server architecture, with significant processing done on a PC client. Further, it went without saying, it had to be Windows-based. With the arrival of the Intel-based 386 computers, the PC had become a viable business machine, capable of exploiting the level of maturity and robustness of Windows 3.1, which had only come to market earlier that year. With these two design points as givens, the only real technology decisions to make were what commercial database to use, and what 4th generation language development tool to employ in writing the software.

We considered the Oracle relational database, but it was simply too expensive for the price point we needed to hit. IBM's DB2 was out of contention because it didn't run on Windows, which was also why we nixed Progress at the time, for it wasn't yet available in Windows either. Informix was Windows-based, but, like Oracle, was too expensive. Paradox, FoxPro, and Access met the Windows and the price issues, but we didn't deem them robust enough for mission critical deployment in manufacturing.

The only option that was viable was SQLBase from Gupta Technologies. An advantage of going with Gupta was that we gained a GUI development tool as well, greatly simplifying the task of integration, by having the database and the GUI interface development tool from the same company.

Frederick Brooks, Jr., whom IBM chartered with the responsibility of bringing the System/360 design to fruition, is the author of one of the most cogent books on software programming, the classic *The Mythical Man-Month*. Amid the vast wisdom of that book, he stresses that users highly value information that is presented with rich contextual clues. He states that the "windows" interface, which was originally advanced by the Stanford Research Institute and later by Xerox Park (where Steve Jobs ran across it), was one of the most significant advances in software design for its "conceptual integrity, achieved by adoption of a familiar mental model, the desktop metaphor."[51] He writes that "a clean, elegant programming product must present to each of its users a coherent mental model of the application, of strategies for doing the application, and of the user-interface tactics to be used specifying actions and parameters, *as perceived by the user*" [italics added], as the most important factor in ease of use.[52]

This was, I knew, the single most critical design criteria in building our system, which we had named VISUAL Manufacturing. The old Confucian proverb that a "picture is worth a thousand words" has deep universal and psychological truth precisely because pictures are "visual" metaphors that convey significant meaning with the greatest economy. The driving objectives of our efforts were to exploit not only the visual metaphor that a Windows environment made possible for the first time, but also the powerful intelligence of the human mind to process information visually.

My goal was to have one screen, or perhaps a small set of screens, that was so contextually rich with meaning to the user that anyone could sit down and easily learn to use the system just by looking at the main manufacturing system "window." I wanted that window to provide a clear and unmistakable view of not only what the user was viewing, but what was implied in terms of navigating through the system to ever greater levels of detail, exploiting the power of the mouse click, drop down screens, and drag-and-drop technology—all inherent features of the Windows environment.

The metaphor I struck upon—not too surprisingly, perhaps— was the multi-level bill of material, but with much more contextual richness due, in part to, to a graphical "card" display. The notion of using the multi-level BOM was a direct outgrowth of being encouraged by people such as Bob Sheldon at Anderson Tool and others to resolve the dire problems created by the industry's use of the level-by-level BOM, where each subassembly had its own unique part number, engineered years back because of the memory limitations of early hardware. I wanted a system where there was not only one customer order number that commanded the entire BOM, but also one that also showed all associated production operations in the routing in sequential fashion. Here again, this could be easily displayed by the graphical card metaphor used to present the bill of material. By combining the multi-level BOM with the routings, we essentially could present a "bill-of-manufacturing," one that could be displayed in one window to provide a powerful intuitive orientation to how the system was structured and how to navigate through it. In doing this, we were effectively giving manufacturers what they once had had in the old process sheet that had served them so well and so long—until early software designers took it away from them due to computer hardware constraints in the 1950s. Only now, what they would gain was a tool that was powerfully interactive.

The Windows environment enabled us to create a sense for a visual, three-dimensional presentation of cards that were linked in clearly understood relationship, one to the next. At a glance, the user could "see" both the relationship and the sequence and, by navigating via point-and-click technology, could select a particular card in the sequence that had specific relevance at the moment in order to "drill down" to view underlying supportive data and see the chain of sequence and relevance at each of the various levels beneath the main user window.

The design of this bill of manufacturing "window" in our system was the foundation for the design of the entire system. It embraced and displayed the full potential our system offered to a customer. The design was elegant, yet simple, and came together quickly in only about four or five iterations.

Beyond the design of this initial system window, perhaps the other most significant design question, in terms of Dave's investment of time, had to do with the scheduling system. Finite scheduling was where we had broken new ground and established our key expertise in building ProfitKey Systems, and though Dave had made major contributions to its development, he had not had a hand in actually coding it; that had fallen to Bob Davis. Now the design and coding both were completely on his shoulders.

Much of the system functionally he managed to code in one- to two-week bursts. This was true of order entry, purchasing, inventory control, and costing. He had done these before. But with scheduling, it was different. Scheduling is a complex problem, one of how to model the real world in a digital framework. Availability and use of people, machines, and materials are all challenging entities for how best they might be represented in a digital format. With scheduling, there was perhaps as much thinking time involved as there was coding time. He knew he was not going to build the "perfect" solution in the early iterations, but he wanted to build one that was the best, with the most resilience for future expansion, in the time we had. The complexity of mastering this requirement took nearly a thousandfold increase in time, however, over that required for any of the other pieces of the system.

All the while, we were running the company on a prayer and a fistful of "company" credit cards. Dave was working from home coding manufacturing and I was coding the general ledger component in a small office I'd leased in Hampton, New Hampshire. I was starting to talk to people I wanted to bring aboard when the time was right, and was beginning to think about how to bring this product to market. My daughter, Suzanne, was handling the phones and the myriad details of managing the office and beginning to pull together lists of prospects to go after when we were ready to launch.

In late summer, Suzanne and her husband, Rich Lagoy, introduced me to Ron Ripley, someone they insisted I had to meet. Ron was their neighbor, and as luck would have it, a man of exceptional organizational talent and expertise. Ron had recently sold a small business of his own and was casting about for where to next focus his energies and talent. We met and hit if off together and I hired him. With Ron on board, we began ramping up our marketing effort in earnest, with the hope of making our first sale by the end of the year.

13 The Vision Made Real

The challenge and the mission are to find real solutions to real problems on actual schedules with available resources.

Frederick P. Brooks, Jr., *The Mythical Man-Month*

By late autumn, I was eager to takes the wraps off the product and begin aggressively marketing it. In my mind, by far, the most important milestone of a start-up company is the first sale.

In November, I called together a small group of sales people I had worked with over the years to demo the system, to gain from them, hopefully, measured in their response, a validation of "proof of concept" for what we had been creating. Though it was not yet fully functional, the system was far more than a prototype. Designed to support make-to-order production—long an anomaly, now the growing norm in manufacturing—our package covered the spectrum from product definition in the estimating module, through order entry, scheduling, purchasing, inventory management, production, and shipping, to accounts receivable. I started the demo with the main manufacturing window, where all materials and operations are visually sequenced behind the customer order using the overlapping, chained card metaphor, and walked the group through the building of the order. I used the Windows features of drag and drop, pull down menus, and drill down capabilities, all tied to the simple click of the mouse, to navigate through the scheduling function, posing and answering typical questions that manufacturers have to make every day to effectively perform their jobs. The response from the group was amazement—followed by raw enthusiasm. The salespeople I had invited were ready to leave their present jobs and start knocking on doors.

Not yet, I told them. Not yet, but soon. We needed to pass the litmus test of having our first customer in hand before we started crossing all the other bridges that we hoped would follow. And while I had been confident that Dave and I could pull this off, it was a huge vote of confidence to see the enthusiasm for the product in the small group of colleagues who had come to Hampton for a first look under the covers.

The first product sale in any industry is always the hardest, but this is nowhere as true as in the software business. Software sales, particularly of business systems, are highly reference dependent. People want to know that other people have used it successfully before they take a leap of faith that it will work for them.

We were fortunate in getting a solid lead soon after Thanksgiving. I had gotten word from a local computer store that had sold them hardware that a small job shop up the road in Portsmouth, New Hampshire, was interested in an inventory package. I made an appointment, went up in early December, and made the first call on the manager of MIS. It was not long, however, before he abruptly threw me out of his office. He was outraged, it seemed, that I had priced a ten-user system at $40,000. He wasn't interested in, nor could he see the value of, a system such as ours. He was interested solely in addressing the users' request for a package that could help them manage inventory. There was nothing I could show him that would justify, in his mind, spending more than $3,000.

A couple of days later I got a call from the vice president of finance. He had gotten word of what I had started to lay out to the MIS manager, before being summarily dismissed, and he was curious. He wanted to meet with me. So I made the trip up to Portsmouth again, only this time I got the opportunity to fully demonstrate the product. The vice president of finance was impressed. He had no trouble grasping the value of estimating, order entry, scheduling, and the rest that was in the first release of VISUAL Manufacturing. The fact that we didn't have any references was an issue, but one he was willing to live with. In our favor, there was nothing else like VISUAL Manufacturing available on the market. And he knew it. He also was reassured by the fact that Lilly Software was only a few short miles away, so that if he needed assistance, we were easily available. This was enough to win the deal. In the early part of December, we signed our first software license agreement with Conceptronic, the small job shop in Portsmouth.

Our second and third sales followed quickly in the opening months of 1993. Profile Metalforming of Raymond, New Hampshire, signed in January, and O-A, Inc., a precision machine shop in Agawam, Massachusetts, that did contract work for Pratt & Whitney, signed in February. Profile Metalforming proved an ideal implementation for VISUAL Manufacturing. It had run its shop entirely by manual procedures and had struggled for years in responding to customers calling to inquire about the status of their orders. The people at Profile loved the visibility that VISUAL Manufacturing provided on all work-in-process. They raved about the ease of use, of having all the production information right at their fingertips. They took to the system with glee and became expert users in no time. And they worked closely with us for months to become one of the best reference sites we have ever had. Rich Lagoy came aboard in early spring to head up customer service, working intimately with all of our early customers, helping them to get up and running on the software. My son Michael came aboard soon after in our associate structure, taking responsibility for handling sales for the New England territory.

In quick succession, we signed Jonathan Shoolman as the associate to handle the mid-Atlantic region, working out of Philadelphia; Skip Casamatta, to handle the Midwest region out of Cincinnati; and Bob Washburn in Atlanta to handle the Southern region. I had worked with them at ProfitKey, where they had all been top grossing salesmen. We had personal ties to build on, but beyond that, they were enthused by what they saw and believed that they could easily sell such a system in the market.

By late summer 1993, our lead-generation efforts had tapped such a wellspring of sales opportunities that we could barely keep up with the volume of requests-for-proposals we were feeding to our growing list of associates. It was a problem keeping up—but the kind of problem you love to have. By the fall, the volume of leads was so great we were confident we were riding a winner. Our goal for our first year in business was to do $1 million in sales. It was an auspicious sign to all of us when we met our goal—and exceeded it, closing the year with $1.4 million in cash accruals.

My son Mark joined us the following year. After graduating from college, he had come back to New England to work for Hewlett-Packard in Andover, Massachusetts. He joined us, originally working for Skip Casamatta in the Midwest region, but eventually took over the New England associate organization.

In 1994 we expanded internationally, hiring Anthony Maurno to develop our associate program in Europe. Tony and I had known each other going back to Software International. When we first talked about his coming to work for us, he commented he thought we were being very aggressive expanding Lilly Software Associates (LSA) into the international market in our second year of business. But he saw the potential in the product and was eager to come on board. He did a fabulous job getting VISUAL Manufacturing established in Europe, signing associates in the United Kingdom and on the Continent, and subsequently came back to the States, where he eventually became Senior Vice President of Sales and Marketing, chartered with building our associates program worldwide.

The associate structure is an ideal mechanism for putting feet on the street at the lowest initial cost, but it is not necessarily the easiest structure for managing the growth of a company long term. Tony's genius is being able to support, prod, and motivate what are essentially independent business operators to seek greater achievement, not merely meeting annual goals, but exceeding them to capture market share as quickly as possible. He turned our marketing and sales apparatus into a powerful engine for success.

We followed our first year's $1.4 million performance with a $4 million year in 1994. We came out that year with the first major new release of the system, expanding its breadth to cover not only MTO, but also make-to-stock (MTS), with MRP tied to forecasting; and to offer a solution for the hybrid environment of MTO/MTS.

After much debate within the company, we decided to hold our first user conference in the spring of 1995. We worried that, perhaps, it was premature to attempt this all-important step in the evolution of a small start-up: inviting all your customers to sit in one room and convey their opinions on how well—or poorly—you have addressed their needs. But we were proud of what we had accomplished with our product, and also of what our customers told us they had been able to accomplish using it. We decided to go forward. We set the date, mailed out invitations, and then headed to Orlando, Florida for the four-day event.

The event was, as much as our first sale and our first million-dollar year, a real milestone. Eager—and at the same time, nervous—to have all our customers gathered in one place at one time, we were stunned by the response we got from them. It wasn't as much what they said, and they told us plenty, but rather, it was the spirit that filled the conference rooms where we gathered. They were excited to be associated with Lilly Software's success. And after four days rubbing elbows with over 125 congenial, contented customers, all of us at Lilly Software returned to the routine of our jobs with a real boost of enthusiasm for what we were doing.

We were on a run, no doubt about it. We had managed to successfully bring together the technology, an innovative design, and our years of manufacturing expertise to create a compelling value for manufacturers who were seeking real solutions to real problems. We had achieved something that had never been done, and our mounting success was more powerful than any marketing claim we could boast.

We closed out 1995 with $8 million in sales, surpassing the 500-customer list milestone. The next year, and the year after that, was more of the same. We nearly doubled sales in 1996, and doubled them again in 1997, closing 1997 with $26 million in sales, with our customer list approaching 1,200.

In 1998, *Inc.* magazine recognized Lilly Software Associates for being in the top echelon of its exclusive 500 fastest growing U.S.-based independent companies. *Inc.* ranked us number 102, with a compounded five-year growth rate in excess of 2,200 percent.

As exciting as making the *Inc.* list was, it proved just the beginning of the accolades we began to receive. We were subsequently recognized by Deloitte and Touche as the 27th fastest growing New England technology company. We were also included in *start* magazine's list of "hottest companies" selling into the Microsoft Windows market. And the Governor of New Hampshire signed a special proclamation in 1998, citing our phenomenal run of success as serving as an engine for growth in the state.

The run of success in our first six years in business was indeed heady. In that brief span of years we saw our growth registered in experiencing our first $1+ million year in our first full year in business, to our first $1 million quarter, our first $1 million month, our first $1 million week, right up to our first $1 million day. In the process, we not only had managed to establish Lilly Software as a powerhouse in the Windows world of manufacturing software, but had also helped create two $10+ million companies, two $5 million companies, and several $3 to 4 million companies among the independent associates that represented us. And as of this writing, we are approaching the signing of our 2,000th customer worldwide. VISUAL Manufacturing is available today in English, German, French, Dutch, Spanish, Italian, and Chinese.

Of all the milestones we have passed in our first seven years in business, however, perhaps the one most significant in my mind is the United States patent Dave Layne and I were awarded in July 1998. U.S. patent #5,787,000 was granted to us for our software design for concurrent scheduling of material requirements and operations. The patent citation states that Lilly Software offers a "computerized method for scheduling a plurality of work orders for manufacturing products in a manufacturing process, each work order comprising a set of operations to be performed using a plurality of resources and materials and methods ...."

The patent provides validation for much of what I had been publicly arguing regarding the long-standing shortcoming of manufacturing software and the persistent resistance by the industry to embrace needed change. As an industry, software developers and marketers, educators, and consultants had been complacent to accept long-held assumptions, assumptions that once had some merit, given the limitations of early hardware. Many of these assumptions, from infinite capacity to standard leadtimes to level-by-level computations, had the effect of compounding problems on the factory floor. But all too often, rather than clearly acknowledging the limitations that forced these assumptions on us, we had been an industry willing to promote the most untenable premises with such aplomb it was as if we had come to believe our own marketing.

I knew when Intel delivered its Pentium-class chip and the first Pentium machines were shipped, that we were finally at the threshold of affordable memory and speed to address these problems in the manner long required. As soon as the Pentium machines were shipped in 1994, Dave, Michael, and I sat down to address one of the most intractable problems in the lot: the concurrent consideration of both material availability and resource capacity. In truth, it represents the conundrum of manufacturing: do I have capacity *and* the materials to fulfill a customer order in the time the customer has requested? And if not, what do I have in the way of materials and capacity to satisfy some portion of that and any other competing order? Our algorithm holds that nothing will be scheduled without all the parts being available *and* without all the capacity required to finish the job by the intended due date being available as well. We saw our way to work out the design in a fairly straightforward fashion. That design became the basis for concurrent scheduling of materials and capacity in VISUAL Manufacturing and the basis for the granting of our patent.

The patent gives Lilly Software a clear competitive advantage in the marketplace, but more importantly, in my mind, represents delivery on promises that the software industry has been making to the marketplace for years: that information technology can play a central role in improving the day-to-day operations of the plant, addressing the myriad complexity of issues that manufacturers have long wanted to resolve. It puts an end to the need for software developers to employ elaborate, convoluted explanations in logic and foolish obfuscation of terminology, all of which served only to sell software. This was typically at the expense of making manufacturing executives feel stupid for their failure to grasp what we were talking about at the time we were closing the deal—and irate after they had the software installed and found out it did not work the way they need it to. We were now entering an era where it was patently unnecessary.

With the present class of personal computers, we have reached a point, finally, where a little common sense can go a long way toward solving the real problems at hand.

14 Common Sense Throughput

Common sense is not a simple thing. Instead, it is an immense society of hard-earned practical ideas—of multitudes of life-learned rules and exceptions, dispositions and tendencies, balances and checks.

Marvin Minsky, Professor of AI, MIT

If we came to realize anything about manufacturing in this country in the late 1970s and early 1980s, it was the strategic importance of manufacturing production to the success of the enterprise. Prior to that, manufacturing—what went on inside the four walls of the plant—was the ugly stepchild in the enterprise family, valued as far less important than marketing, sales, and finance. It was a rude awakening when American manufacturers discovered that not-so-distant competitors from the Far East and Europe understood the importance of manufacturing prowess far better than we did. We spent the 1980s and much of the 1990s trying to put the American manufacturing house in order. Much of this focused on applying new technology, particularly information systems, to the basics of making better products faster at less cost—with a growing realization of the importance of customer service.

As we enter the 21st century, high-quality customer service—both to internal and external customers—has become the key point of competitiveness. Quality products and price are still essential, but they have become a "given," the threshold merely to enter the competitive arena. Make-to-order and its modern twin, mass customization, are becoming the norm in all industries, not just in traditional job shops. The emphasis on customer service is taking deep hold in our organizations. In a recent *Information Week* poll of 300 companies, 69 percent of the IT executives surveyed stated that they were "very committed" to serving external customers, up from 42 percent only a year before. Yet, while interest is up, we still have a long way to go: less than a third of the surveyed respondents in the poll met *Information Week's* criteria for what it termed a "customer-centric" operation.

We have the technology—finally—that is capable of achieving much of the long-promised vision heralded by industry marketers thirty-five years ago, when we first began to dream and scheme of applying computers to help manufacturers accomplish more with greater efficiency and less cost. It is tempting, however, to continue to be lured by ever-increasing promises of the riches available just on the cusp of tomorrow, as the relentless advance in computer technology continues to push the envelope on what is imaginable.

A word of caution to manufacturers: don't be seduced to wait. Don't be blinded by the dazzle of promises of what is yet to come, for the wealth of what is possible today is more than enough to justify the effort to extract it with what is at hand. I have come to refer to this opportunity—one that far too many manufacturers are continuing to overlook—as Common Sense Throughput. It is a notion I struck upon some years back, when I realized that we finally had passed over the threshold that had been holding us back for so long. Dr. Eli Goldratt, the well-known educator and business consultant, had also embraced and made accessible the concept of throughput in his landmark business novel, *The Goal*. I started calling it Common Sense Throughput to make the point that we no longer have to look to some distant tomorrow to achieve dynamic gains in throughput, productivity, and profitability that would bring joy to the most hardened CEO or corporation finance officer.

Common sense, however, is not all that common. This is precisely why it is valued. We call it common, for when we see it for what it is, we are always surprised that we had not thought of it sooner.

Such is the case with manufacturing throughput. People think of throughput in different ways, but in point of fact, it is the most critical turnstile for determining profitability. Of course, balance sheet profit-and-loss calculations routinely have other associated elements tethered to them. But at the end of the day, it is throughput that sums things more quickly on the bottom line than anything else under our control.

Working to order, if you made more today than you did yesterday—without any discernible increase in costs—you have advanced your game. If not, you have started a trend that ought to light a fire in the company, a fire with such intensity that a remedy is everyone's concern.

That is common sense.

But how common is it in our organizations today? Have we become so complex, so sophisticated in what we do that we can't see this? Or worse, don't find merit in it any longer?

In truth, in actual practice, it not that common. But that is not to say that it doesn't exist, that it isn't being practiced by forward viewing—you might even say *revolutionary*—production managers, given the prevailing mindset that continues to shackle management thinking in many quarters. We are encouraged, though, by a growing number of cases where top management has been willing to go back and question old assumptions, apply new tools, and achieve startling results.

Instrument Technology, Inc. (ITI) is a privately held company located in Westfield, Massachusetts, at the edge of the Berkshires. ITI manufactures remote viewing scopes used in medical and industrial applications, ranging from advanced surgical procedures to inspection and test of jet engines. The company was founded in the 1960s by Donald Carignan and partners to design custom remote viewing technology for the nuclear industry, where access to critical areas of the radioactive reactors was severely restricted. Initially focused exclusively on design, ITI subsequently brought manufacturing in house through acquisition of one of the key job shops with which it frequently contracted. With the downturn in the nuclear industry in the early 1970s, Carignan took sole ownership of the company and began to diversify the product line into aerospace and then in the 1980s into the emerging medical instruments market.

Today, ITI has a product catalog with over 2,000 items, nearly all make-to-order. The prices of scopes range from $1,000 upward to $30,000. The medical instruments business represents the greatest area of growth. In medical instruments, ITI works with a handful of original equipment manufacturers (OEMs) with whom they contract, supplying the scope technology that is integrated into more comprehensive systems. The company continues to do most of its business in North America, but has a growing presence in the international market as well. Management of the company has been passed to the next generation, Donald Carignan's three adult children, Dawn Thomas and Gregory and Jeffrey Carignan.

The manufacturing side of the business is composed of both a machine shop and an assembly area. The machine shop takes various metal stock, including bar, tube, and flat sheet material, and converts it to component parts through a series of milling, grinding, drilling, and turning operations.

These items are then kitted to orders and passed over to the assembly area for final build. The shop tends to work to simple, single-level bills of material and is typically capacity constrained; that is assembly works to multi-level BOMs and is material constrained.

In the early 1990s, the growth of the volume of business being put through the facility strained the information system in place beyond its ability to cope. The system was classic "sneaker-net," with four isolated PCs each handling either order entry, purchasing, accounting, or manufacturing. Work-in-process was out of control; assumptions about material availability were creating bottlenecks in assembly; and there was no easy way to address customer inquiries regarding the status of orders without considerable effort and delay.

We worked with the management from ITI soon after we first brought our Windows-based solution to market. ITI had looked at numerous other packages, from midrange solutions down through the welter of PC-based products then available, but hadn't found anything that provided them the breadth of functionality they needed in a user-friendly package. We were able to provide them the ability to track actual—not standard—cost and apply the power of finite capacity constraint scheduling to production. And this was in a highly intuitive, easy-to-use Windows application on their desktops.

ITI began attacking their production problems using common sense. The production management team began to meet regularly to assess throughput of operations in previous periods. They began to use the information the system provided to determine why the figures changed. They began to strategize what they could do to effectively boost the numbers.

Since then throughput in the plant has gone up almost 160 percent—without a corresponding increase in headcount. All prior efforts to boost throughput had achieved only modest upturns, largely as a result of adding more people.

What they did was common sense.

But why is it so uncommon?

Let's break it down. Let's look at the simple elements that makeup throughput as a factor of production, but more importantly, as a leading indicator of the profitability of a company.

The real significance of throughput is often blurred or obscured. It loses some of its sharpness as a measurement of general health of operations when coupled with other measurements, typically those employed in a "balance sheet" approach to calculating revenues versus operating expenses. *Cost of goods sold* and *outside services* are commonly rolled into that equation. And as a result, the potential leverage of using throughput as a point of focus for boosting profitability is minimized.

The real value of throughput should be determined by subtracting the materials and outside services from the selling price. Throughput is calculated only for products or jobs that are sold or for which we have a customer's purchase order.

At first glance, the common sense of this might be hard to reckon. What about labor? What about overhead? What about inventoried finished goods?

Good questions. But think about them. Where is the leverage in labor and overhead and unsold finished goods? Can a CEO increase the utilization of the facility without increasing labor and overhead?

If the throughput can be increased without increasing labor and overhead, you have effectively boosted not only throughput, but more importantly, profitability. You've squeezed more juice from the fruit at no additional cost—and it all goes to the bottom line.

By stripping out labor and overhead, you are forced to focus on what you *can* effectively control to increase profitability. It enables you to focus on those things that inhibit—or more strategically, *accelerate*—throughput. Make more goods in the same amount of time without increasing costs and the additional throughput is all profit. It is common sense.

The old adage that there are no second acts in life is convincingly dispelled by the success of Dearborn Precision Tubular Products, Inc. Located in one of the most unlikely sites for an industrial operation, nestled against the spectacular backdrop of the White Mountains in western Maine, the company is a specialty job shop dedicated to what is known as deep hole or trepan drilling of high alloy metals. Located in Fryeburg, Maine, Dearborn Precision Tubular was an outgrowth of Howard Dearborn's interest in keeping his mind active and his hand involved in the field of mechanical engineering after retiring and moving from the Midwest to the beauty and serenity of the Maine woods. What started out as a tinkering in retirement grew into a viable business, one that surpassed the size and scope of the manufacturing firm he had founded in Cleveland, that he had turned over to his son to manage.

Dearborn Precision Tubular manufactures component parts primarily for the nuclear propulsion, aerospace, and oil and gas industries. A brilliant, self-taught mechanical engineer, Howard Dearborn designed and built many of the more critical machining centers for mastering the exacting precision of deep hole drilling, holes from .062 to 8 inches in diameter with depths up to 40 feet and tolerances of less than 1/1000 of an inch. Machined components are used in medical equipment, jet engines, and drilling heads on oilrigs operating in the most hostile and demanding environments imaginable.

What started as a simple aspiration to keep busy grew through the 1970s and 1980s into a viable job shop employing upward of a hundred people by 1990. As the business grew, so did the nature of the business, as its industrial customers looked increasingly for greater service as they began to outsource a greater proportion of overall production. What started as a specialty in deep hole drilling took on the full measure of a machine shop as Dearborn added turning, grinding, and milling to its in-house operations as the complexity of the components and the scope of services expanded to meet its customers' needs.

As the complexity of the components grew and the number of operations required to produce them expanded, the complexity of managing production grew exponentially as well. With service and quality key differentiators for determining long-term viability as a component supplier, Dearborn was faced with either upgrading its production management system or risking the loss of its competitive edge.

The company had been scheduling the shop by a manual process using a magnetic scheduling wallboard. The board had started out as an eight-foot panel, but as customer orders and machines had been added through the 1980s, the board grew panel by panel until it stretched 24 feet across the wall of the scheduling area. This was, in many ways, not unlike what Jesse Jones had devised at Associated Machine in Miami using what he called his abacus-like "Chinese computer system." At Dearborn, delivery to promised ship dates came to be problematic. Having to manipulate an increasing number of orders through a growing maze of operations became impossible, in terms of having visibility to the impact of the ceaseless stream of changes inherent in running a shop at peak capacity day to day. The company was losing control of its most vital competitive feature—its shop capacity—but not for want of effort. It had simply become too complex to manage.

Bill Findeisen, general manager, chartered a small team to begin a review of possible solutions to the problem. This effort gained critical momentum in early 1998, when Dearborn asked Dave Hague, a manufacturing systems consultant, and Rob Newton, an accounting systems professional, to join the company full time. The two of them, along with Ed Hermance, who had long been managing the demanding task of scheduling the shop, began talking to us. Though they had been assessing other solutions at the time, they were keenly interested in the "what if" simulation capability of the finite capacity scheduling function of the system as we designed it to work. They licensed the software in short order, laid out an implementation and training plan, and were up and live on the system in eleven weeks.

At any given time, there are typically 200 to 250 orders working their way through the shop. Product lead times vary from one or two days to 16 or 18 weeks. With 200-plus orders, there are typically 4,500 operations to schedule to ensure that the jobs get done and shipped as promised to the customer. Before Dearborn began working with us, there was no way for management to assess the impact of an action—from entering a new order to changing an existing order or product configuration already in production—on the ability of the plant to meet existing customer commitments.

We provided Dearborn with a solution to this problem, one that was simple, fast, and comprehensive. By employing the use of multiple schedules in the system—a standard schedule reflecting current work-in-process, and various other iterations for planning and dynamic "what if" simulations—management gained the ability to quickly assess various scenarios, whereby they could measure the trade-offs involved.

The software permits both backward and forward scheduling. Backward scheduling is typically used simply to complete an order "just in time" to meet the due date, whereby the system schedules the operations required to meet the due date by filling all available "holes" in the existing load on the shop. Forward scheduling is also pegged to the due date, but enables one to start later than what is indicated on the backward pass, and schedules all the operation start dates required (given operation run times) to execute the order with priority in the system. Classically, forward scheduling creates significantly more contention among orders waiting in queue at the various work centers.

The system provides a Contention Inquiry window that shows the contention, measured in number of days' wait, among the top resources in contention given any particular "what if" simulation of the schedule. Identifying contention of resources is, in effect, the means for identifying "bottlenecks." The contention window graphically displays the top ten bottlenecks given the scenario the scheduler is considering. The first time Dearborn ran its scheduling through the system, the resource contention was in excess of 400 days.

Contention is not, by definition, an evil. The real issue is how you manage it. We enabled Dearborn to modify the schedule, giving them the visibility and intelligence of what the impact of changes were, such that they was able to bring that 400-day contention down to two days with only a few iterations of change to the plan.

The Contention Inquiry window, to be employed effectively, requires someone who knows how the shop works. But what the system provides—and what classically has been missing before this—is a tool that is as sophisticated and powerful, and yet easy to use, to help the scheduler balance the schedule to achieve the greatest good for the company while also meeting the commitments to the customer.

In the short term, Ed Hermances' job as scheduler became much more complex; he had to learn how to react to the truth of the changes he was making in the schedule. But in the long term, his job became much easier after he rounded the learning curve, for his knowledge of how the shop operated was now matched to a tool that enabled him to wring as much proficiency and productivity from the shop as possible. As an added benefit, he was liberated from the onerous task of manually maintaining the magnetic scheduling wallboard.

In a matter of weeks after starting to hold twice-weekly Throughput meetings with the management team at Dearborn, Hermance was able to significantly boost throughput in the shop, while reducing contention of resources by as much as 70%. Overall shipment to quoted delivery dates has improved a minimum of nine days.

In manufacturing, that which impacts throughput is a *constraint*. When the influence is negative, it is called a *bottleneck*. Eliminate or reduce the bottleneck and you boost throughput. With the caveat, of course, that *everything else remains equal.*

But as anyone who has worked in production management well knows, bottlenecks rarely remain stationary. As you tweak and gain efficiency in one area, or your mix of customer orders changes, typically you'll witness the emergence of a new bottleneck somewhere else.

That being the case, knowing where your bottlenecks are—and where they are likely to move—is a powerful bit of intelligence in effectively managing production to boost throughput. And profits.

Imagine, if you will, a manufacturing company with $5 million in revenues. Expenses, comprising $3 million in labor and overhead and $1.5 million in materials and outside services, total $4,500,000. In this scenario, therefore, gross profits amount to $500,000, or 10 percent of sales.

Now lets consider a 5 percent increase in sales—$250,000—and by applying Common Sense Throughput methodology, the company is able to maintain labor and overhead without any appreciable increase. Materials and outside services, however, increase by $75,000, netting a new overall total in material and outside services equaling $1,575,000.

Running the math on this, gross profits are now $675,000, or 13 percent of sales. Total throughput, therefore, has increased to $3,675,000. This equals an increase in throughput of $175,000. The resulting increase in throughput goes directly to the bottom line. So, with the original 5 percent increase in sales, we were able to boost profits an impressive 35 percent.

And how did we achieve this? By managing capacity such that we enabled more throughput with existing resources by closely monitoring and controlling critical bottlenecks.

But how many CEOs meet with key production managers on a daily basis to scrutinize the behavior of yesterday's bottleneck? To explore what they are going to do to manage it better today, or to manage the resultant new bottlenecks? To identify where bottlenecks are most likely to appear over the next two to three weeks, and what they are going to do about that new set of problems?

If you are interested in boosting profits, doesn't it make sense to focus on these issues?

At ITI in Westfield, Massachusetts, management uses the software strategically to confirm where bottlenecks reside on the plant floor. In one instance, a new CNC machine soon proved to be a bottleneck. Information that management was able to glean from the system helped them identify that the problem was not what the machine could not do, but that it could do so much more. Engineering was now designing parts the company had never been able to machine. In addition to this intelligence, we helped them segment the tasks being assigned to the new machine and identify a series of operations that could be grouped and performed on a much simpler machine. The system helped management justify the purchase of a dedicated resource for offloading some of the work, effectively resolving the bottleneck issue.

Because ITI is a small company, the allocation of human resources is also a critical management function. The system helped management balance the assignment of one skilled worker who had been moved from the shop floor to help with quality control, when quality control was proving to be the bottleneck in shipping orders. This move, however, had the effect of forcing the bottleneck back into the shop. Now, with aid of the system, ITI can anticipate when to move the crosstrained worker from one area to the other to better manage the overall flow of jobs through the plant.

With all the information that is gathered on operations, from worker hours to job duration to efficiency, management is better able to manage job quoting and new product introductions, because they have intelligence on the dynamic nature of ramping up from design through peak efficiency.

This provides ITI a better handle on quoting price and delivery on the front end; while the software also aids them in keeping the commitments they have made.

The net result at ITI is that they are doing more business with less effort and cost, while doing a much better job of meeting the expectations of their customers.

Much like ITI, Dearborn looks forward to using the software to aid them in making strategic capital investments as well as to scheduling production through the plant. Capital investments, in either new machines or new operators, must be strategically managed in order to match the increase in business with increased capacity. Dearborn expects to be able to better manage this tricky task in terms of timing, as well as to pinpoint precisely what capacity elements need to be added to meet the exact needs of the business.

Like ITI, Dearborn is singularly focused on maintaining the highest quality commitment to customer service. Its investment in a true state-of-the-art software system was cost justified on this one issue. It has well proven to be a versatile, powerful tool for helping Dearborn achieve its goal and better position itself as it enters the 21st century.

Common Sense Throughput is a methodology based on simplicity. It is founded on the notion that it is the CEO's job—a highly strategic job—to know what is going on in his plant on a daily basis. The CEO can achieve this by committing as little as fifteen minutes each day with key staff members to analyze the organization's bottlenecks. The purpose of the meeting is to discuss the current state of the bottlenecks and take action on them in order to prevent a negative impact on profitability. The staff members should arrive at the meeting prepared to discuss three main areas of focus: yesterday's bottleneck—how and why it performed the way it did against plan; today's anticipated bottleneck and how they can increase capacity to increase throughput; and the anticipated bottlenecks looking over the next three weeks, the reasons for them, and what can be done to achieve the greatest utilization of each one.

So the question becomes: *how* can we best identify our bottlenecks in production? And with enough responsiveness to proactively effect a positive impact? By employing Common Sense Throughput. And by using information technology that is designed specifically to address the fundamentals behind the methodology.

The phenomenal success that Lilly Software Associates has achieved is one indicator of the validity of the premise. But more significantly, we believe the testament of truth lies in what manufacturers have been able to achieve in their operations and in their balance books employing the Common Sense Throughput methodology. Their success speaks far greater volumes about the validity of Common Sense Throughput than anything we could say.

Abrasive Technology, Inc. (ATI) lies at the crossroads of traditional American manufacturing and the globally wired virtual manufacturing enterprise of the future. Headquartered in Westerville, Ohio, in suburban Columbus, its goal is to be the dominant provider of industrial and medical/optical superabrasive material removal tools through a combination of providing the highest quality products in the shortest time and employing the latest in information technology.

ATI comprises a family of eleven different corporations located in the United States, Canada, Spain, Singapore, and the United Kingdom. Ranked among the top 10 of the 350-some suppliers of superabrasive tools to industrial companies in the United States, it occupies the special niche of currently being well below the top three manufacturers, but well above the remaining pack that fills the bottom of the market. Small by comparison to the largest players, the company is in no way small in its ambition.

Founded in 1971 by Loyal M. Peterman, Jr., and partners, ATI specialized in providing diamond abrasive products used in cutting and finishing a variety of surfaces and materials, from the heat shield of NASA's space shuttles that carry astronauts into space to drills used to eradicate tooth decay from inside the human mouth. Peterman was educated and trained as a mechanical engineer, but has proved to be one of those exceptions in engineering, crossing over successfully to master the nuance of entrepreneurial endeavors.

Today, ATI operates 11 plants in five countries. Much of the growth of ATI has come in the last thirteen years through a series of strategic acquisitions aimed at adding breadth and depth to its product line as it was also adding reach to its vision to become an international supplier. ATI has over 5,000 customers today, with 80 percent of its business currently based in the United States, but has plans to grow its business in Europe and Asia over the next few years to equal what it does in North America.

In 1996, prompted in part by the Y2K issue, ATI began the search process to replace its existing production information system. The company had been system-savvy for fifteen years, migrating through a series of solutions including an IBM System 36, then a System 38, before migrating in 1987 to the HP

3000, using a software package tailored to the job shop environment. It had growing concerns, however, about the long-term viability of this system to address its requirements, and found in the Y2K issue incentive to upgrade again to a new server architecture and a new software package.

In all, ATI evaluated 20 to 30 packages through the summer of 1996. They liked what they saw in our system and invited us back for a second round. The demo was scheduled for a Saturday, and Peterman had packed the conference room with 50 associates ranging from senior associates to individual distributors invited to make all their concerns known and have issues addressed. The demo was based on real data from ATI and lasted four hours.

The demo validated all the critical points on their needs list, most significantly that they could build estimates and generate customer orders directly from the estimates. The audience was impressed with the overall visual nature of the system, its ease of use, and the straightforward navigation it provided through all the levels of detailed inquiry they requested. ATI signed the licensing agreement October 1, with the objective to go live with 65 users on January 1, 1997.

The project scope and system implementation was aggressive and not without its problems, caused primarily by glitches in the interface between the application software and the relational database. Dave Layne was dispatched to Columbus at the eleventh hour to resolve last-minute critical issues and got them through the rough spots to full operational use.

ATI manages a 40,000 part item master in the building of its products. Make-to-order production at ATI typically includes a combination of routings through the machine shop and then through the process area where the bonding techniques are applied to adhere diamond chips to the metal surfaces of the machined components. Routings typically include three to seven steps, with a production leadtime that varies from three days to four weeks.

Though ATI is currently using only the scheduling feature for its medical products group stock product line, the system provides the company with clear visibility to the status of all customer orders in production. Even using scheduling in a limited fashion, the system has improved overall throughput and the timely resolution of contention of orders. It is a part of ATI's overall strategic vision to bring the entire product line under VISUAL Manufacturing finite constraint-based scheduling in the near future.

The software also serves as an information engine for an elaborate global information system that pumps updates to all of its sites and all of its sales representatives and customers around the world via Lotus Notes. Customized "alarm buttons" added to the system highlight the arrival of critical notices such that the 145-plus global representatives, as well as all the local users, have notice of the latest postings of global messages. This feature provides real-time intelligence for building estimates and quoting and promising delivery, critical in continually seeking improvements for speeding the process of getting customer orders into production.

Additionally, ATI has built an extranet for enabling customers to access the status of orders online. Customers can access open orders, price quotes, invoices, and the status of what has been shipped 24 hours a day, improving the quality of customer service while eliminating the overhead of having to respond to customer queries by phone.

Overall, ATI has seen a general increase in operations productivity of 40 percent. This includes decreasing the time of order entry from days to hours by employing a completely paperless process. There has also been an increase in performance to schedule, with shipping cycle times being improved by as much as 50 percent.

Though Peterman admits that ATI is still classified as "small" by industry standards, the company is anything but small in the vision it has for the future, one he says is completely dependent on information systems technology, with our system serving as the central hub. In measurable terms, ATI has set as its objective to drive the current industry standard for customer delivery from four weeks down to two—and with an ultimate goal of getting it down to 24 hours.

Peterman is a visionary, an engineer, and a builder, with his ideas for the future of his company well under construction. His vision calls for a system-centric, virtual, global manufacturing enterprise. He already has both the information backbone—VISUAL Manufacturing—and the WANS communication infrastructure in place. The vision plans for the global system to provide data, voice, and video transmission, such that orders, work instructions, and even CNC programs can be transmitted real time to anywhere in the world.

Peterman's plan is to put in place a fault-tolerant process and product business model, where every work center in each of the distributed manufacturing facilities can essentially serve as a routing linked with any other work center anywhere else in the world, based on availability of materials, capacity of resources, and the ability to meet the requirements of the customer in the shortest possible time. Peterman stresses unequivocally that scheduling of production—provided by the software—is key to the whole vision of his virtual global manufacturing enterprise.

Peterman's vision is eminently possible. For us at Lilly Software, it is exciting to consider the prospects of being integral in such a strategic implementation. And while the basic functionality is available in the current release of our product, there are some pieces to the technological architecture that still must be added. The good news is that we have already been long at work pulling these pieces together.

Lilly Software is as interested in the future of manufacturing—in the vision of a virtual global manufacturing enterprise—as Loyal Peterman. Like Peterman, we have both the roadmap and the vehicle under construction for how best to support a globally distributed enterprise from an information systems perspective. The concept of globally distributed work cells is strikingly similar to how we envision globally distributed software applications: the job required is performed at the site where it makes the most sense for it to be performed.

Today, this is far more than just a concept. It is embodied in a new generation of software applications that preserves our deep, long-standing domain expertise in developing manufacturing software and employs the latest in leading-edge technology for how to best deploy it. It is both a vision and a product for supporting how manufacturers will operate in the 21st century.

15 Convergence for the 21st Century

Many people expect advances in artificial intelligence to provide the revolutionary breakthrough that will give order-of-magnitude gains in software productivity and quality. I do not.

... Many students of the art hold out more hope for object-oriented programming than for any other technical fad of the day. I am among them ....

Frederick Brooks, Jr., *The Mythical Man-Month*

Nowhere is change as constant, perhaps, as in the field of information technology. This is abidingly true of software. Having been both the victim and benefactor of the rapidity of the sea of change that sweeps the industry every few years, I have become increasingly alert for the subtle—and sometimes not so subtle—shifts that can mark the beginning sweep that can come seemingly in the blink of an eye, stranding many, while at the same time providing others almost unlimited potential.

At the Lilly Software Associates sales meeting in August of 1997, I announced to the company that Dave Layne was no longer working on VISUAL Manufacturing. I could not have made a more dramatic announcement to the group than if I had declared a mile-wide meteor was to strike the East Coast in an hour's time. Dave's contribution to VISUAL Manufacturing goes without saying and is recognized within the organization as well as widely among our customers. The momentary silence was deafening. And then the protests erupted.

I was rather circumspect in explaining why. In truth, we were putting Dave to work on a stealth project. For those in the know, a group numbering less than a handful, it was understood that the objective was no less than the strategic reinvention of our product line. Dave and a small, select team were going into "deep cover," completely isolated from day-to-day operations at a secret remote site. Only three people would have the telephone number for reaching him: Dave's wife, Ron Ripley, and me.

The reason for this move was our assessment that the industry was on the verge of undergoing another sea of change in the technological base to software development. If this were the case, and we strongly suspected it was, we wanted to avoid the risk of imperiling our growth by meeting the challenge that we anticipated was on the horizon for everyone in the industry.

Dave had come to me some months before concerned for the future of the company. It was not that he had stumbled upon a well-kept secret as much as that he had come to appreciate the true significance of reports that were increasingly in the news. Not too surprisingly, Microsoft was the hand that was stirring the pot.

Microsoft, like many software companies, was actively interested in the promise of object-oriented technology. Object-oriented technology (OOT) is a fairly arcane field, but in practical terms, it promises reusability of software code. It does this via the means, or architecture, by which software can be designed and built using object technology. To a software company, reusability means increased productivity, faster time to market, and more reliable software, among other things. Microsoft—as it always wants to do—was putting its own unique stamp on OOT with something it was then calling the Component Object Model, or COM. COM was, in essence, a specific design, or architecture, for how to engineer object-oriented software—the Microsoft way.

Dave was extremely sensitive, as was I, to something with as much potential to create change—a nice word for upheaval—in the software world. Our history at ProfitKey, getting caught unprepared for the tsunami the explosive rise of Windows created, made us keenly alert to anything that threatened our viability in that lightening-strike fashion.

Our objective in putting a handful of programmers to work on a special project was to explore the emerging new technologies and architecture behind COM. The project was research *and* development; the goal was not merely to study the new technology, but to use it to develop a new product that could augment our current market focus. Dave, Bob Davis, and two or three others began meeting and working integrally with Dave Webster, a Microsoft consultant we brought in to get our team on a fast development track. We housed the team in a nondescript storefront, in a small, nearly deserted strip mall several miles from our main office in Hampton. Most people inside Lilly Software thought Dave and his team were working on a new distribution software package, even though the secrecy of the project seemed disproportionate to such a task.

COM subsequently morphed into DCOM (for distributed component object model) as Microsoft continued to tweak the model. DCOM has since grown into Microsoft's Distributed InterNet Architecture, known by yet another ubiquitous three-letter acronym, DNA. If nothing else, the speed at which acronyms come and go in the industry speaks to the challenge of technology companies, not to mention end users, faced with the task of keeping abreast of potentially disruptive paradigm shifts.

DNA, as it is currently defined, is more a set, or group of technologies, methods, and design principles than it is a single monolithic construct. From a practical standpoint, it provides a framework for separating the various, basic elements of application software programs into three key elemental components: the presentation, or user interface; the application, or business logic layer; and the data storage layer. By separating the three into three distinct elements rather than having them intertwined in one giant monolithic wrap of program code, DNA provides the potential to mix and match, or "plug and play," with greater freedom of choice at an increasingly refined level of granularity of elements.

In more technical terms, DNA—which is the architecture Microsoft is using to build all current generation products—permits software engineers to distribute the user interface, the business objects (or application logic), and data storage across multiple computers to permit the greatest flexibility in system configuration, uptime, and runtime performance in a complex networked computing environment. The key benefits include increased ease of use; increased process improvement flexibility; faster, cheaper application enhancement, with less effort and error; faster, more reliable data retrieval; and faster, cheaper, easier overall system component integration.

While these benefits are significant, representing an order-of-magnitude improvement heretofore unavailable in software development, it is not by any means a simple technology and design technique to master. It presents a new multidimensional field in which conceptual design takes place. Components must be designed to fit into multidimensional schemata. In technical terms, there is an exponential increase in the pathways in which executable commands can flow. The degree of difficulty to master the rigors of the discipline is also increased by an order-of-magnitude never before witnessed in software development.

In real world terms, what we were required to do was to stay vigilant to the ever-emerging evolutionary change that was moving through the software development world, even as we set about the task of riding its wave. This is much easier said that done. It requires keeping an eye on the way forward, while maintaining keen peripheral attention to prevent some new crosscurrent from sweeping us under.

This forced Dave Layne and his group to constantly rethink what they were doing in light of every new nuance of new technologies they were attempting to master. In the early phase of every technological change, things, as a rule, change abruptly, because the technology or group of technologies that serves as the *prime mover* is, in truth, only at the beta level of development. Only once technologies begin to solidify, which is typically when they move from beta into the open market and are being adopted for broad production use, does the rapidity of change begin to slow as the core begins to stabilize. This constancy of rethinking leads inevitably to having to discard some of the ideas that have been put into design and to starting again with a fresh rewrite of the code.

Rethink, discard, rewrite. This is the rule for attempting to stay on the leading edge of a major shift in technology. You're forced by necessity to move forward even though you know that there are weak points to the crest you are riding. When the technology shifts and reforms suddenly, you have to be ready to refocus and reallocate resources to recapture your momentum. This can only be achieved by brute force of will. The only other option is to lose the momentum, the competitive advantage of being the leader as you get swept under, run over, and jettisoned aside.

This was the case with Microsoft's evolutionary path from COM to DCOM to DNA. The introduction of the DNA model marked a significant point for reassessment. We decided it was too important to ignore, that it merited serious rethinking of the work we had already completed, as well as that in which we were engaged at the moment. As it turns out, we determined that DNA was a good strategy. The problem was that acknowledging this created a major point of incompatibility in the business objects and the data objects we had already engineered in our next-generation development effort. To be successful, however, we knew that we could not tie ourselves to an idea and a way of doing things simply because we had earlier decided it was the right way to go.

When you are in the exploration stage of adopting new technologies and new paradigms, not only must you rethink guiding principles, but you must also explore all the various paths and side channels that present themselves as a natural outgrowth of the work that is moving you forward. It is difficult to do this successfully by sheer selectivity. You have to at least poke around to the point that you can determine that a fork in the road will lead you to a dead end, rather than risk the chance you are missing something major that lies ahead on that path for lack of willingness, determination, or courage. We learned this in investigating the merits of DNA, based in part on its predecessors, both COM and DCOM. We feared that at first we were going to have to retrace our path entirely and discard much, if not all, of what we had done. Instead, we validated the principle of separating business objects, data objects, and user interface layers. We didn't have to do much to the user interface, but we did have to go back and rethink, discard, and rewrite both the business objects and data objects in our design.

The "stateless" quality of business and data objects inherent to DNA was contrary to what we developed in VISUAL Jobshop, our exploratory next-generation effort. But we found that rather than throwing everything away, we could select out the best of what we had done and then rewrite underneath the conceptual logic.

Microsoft, in its own exploration of the new paradigm, had determined that "stateless" objects permitted much greater performance in a widely deployed enterprise system than when objects retained memory of the previous processing call made against them (the quality and condition of possessing a "state"). DNA discouraged this, so we had to go back and rework those business and data objects we elected to bring forward.

Another point for rethinking, which we came to feel compelled to make primarily for time-to-market reasons, was the feature permitting users to customize the user interface by changing the code. Dave Layne had initially felt very strongly that this was a feature he wanted to give the user, not only the new users who would be taking delivery of VISUAL Jobshop, but, also those veteran users who would eventually be receiving the reworking of our entire product line. What Dave discovered was that the task of doing this was much more rigorously demanding than expected, netting a significant drop in anticipated productivity in the team. The concept was suspended in the active development in order to keep to the original target date for taking the initial next-generation product through beta to general market availability.

Rethink, discard, rewrite. At the root, you are always testing previous assumptions. In my long history in the industry, I have learned it is being blind to the consequences of hanging on too long to old assumptions that is so detrimental to true advance of the technology and, ultimately, the strategic benefit of technology applied to production and business problems.

This was as true for the emergence of the new technology paradigm embodied in object oriented technology, as it is proving in the emergence of the new business paradigm embodied in the collective arena we have come to refer to as *e-commerce.* Again, the challenge for software developers, indeed for all companies doing business in the world today, is to rethink, discard, and recraft what you have been doing. Traditions are honored because they provide a sense of security and reliability, but they can be killers in business operations just as in the advance of information technology. The advent of e-commerce is totally technology driven, brought on by the ingenuity and willingness to perceive and embrace change. It is interesting to note, moreover, that the points of merit underlying the success of OOT to the development of our next-generation product have all been validated again in our movement to embrace the coming e-world in the rethinking and adaptation of the design of our software to accommodate e-commerce, e-collaboration, and e-speed for our customers to be able to enter the e-world.

Object technology enables software developers to more readily adopt and embed e-commerce functionality in their software for the simple reason that OOT-based systems are not monolithic, but are comprised of thousands of tiny modules, if you will, each with the business and data processing logic contained together, so that you can rapidly extend functionality by adding new objects to the set rather than by having to start over entirely at the beginning once again.

E-commerce represents a new business paradigm, one that embraces seamless integration of front office tasks with product development and production and distribution execution requirements, linking the path from taking a customer order, to checking availability and managing the status of production, on through the creation of the warehouse manifest for wave picking, staging, shipment, and delivery tracking.

The importance of end-to-end supply chain execution is becoming increasingly important because of the speed and ease the Internet gives the customer to transact business. The biggest winners in the new e-world are e-customers and the companies that can best meet their requirements. The proliferation of the Internet as a channel for business empowers customers to quickly and easily compare prices, quality, and delivery schedules. On the demand side, the customer is rapidly coming to the point of wanting it *now*, and the entire structure of global business is being revamped to create Web storefronts in order not to miss out on a single opportunity for a successful transaction.

On the supply side, the Internet is enabling companies from the Fortune 500 to the smallest job shop to expand the breadth of their storefronts to global proportions, crossing geographic borders and barriers of language and time in a heartbeat. The first day that Abrasive Technologies, Inc, in Westerville in suburban Columbus, Ohio, opened its Website, it received an order from a new customer they had never heard of. ATI, like countless companies everywhere, is now open for business around the clock, 365 days of the year.

The Internet is also encouraging closer collaboration of all partners in the chain, permitting the faster exchange of information on availability, cost, quality, and delivery, as well as exponentially expanding information on their customers, the competition, and their industry. It is enabling companies to buy material and products at a lower cost. And it can be done as fast as imaginably possible.

The world of business is powered by the reciprocal relationship formed between those who want to buy something and those who want to sell. Sellers are creating or joining trading exchanges to facilitate cash transaction and to stay competitive, while buyers are shopping harder with less effort, delay, and capital outlay.

Online business growth, according to Forrester Research, has been quadrupling each year over the past several years, and revenue growth is projected to top $1.55 trillion in 2003. And e-business is not simply a transfer of revenues out of the old economy into the new economy, but a revenue generator of proportions formerly unseen.

Talk of changing paradigms in business and technology is vastly different than talking about throughput. But it is essential to grasp the significance between the two. For one fact is becoming important to every company wanting to play on the competitive stage not only of today, but also in the future: slimmer margins, that is, less profitability per transaction, is unavoidable. And this makes execution resulting in greater throughput the key.

True, manufacturers will be able to buy cheaper and work more collaboratively with partners because of the Internet. They'll be able to significantly reduce transaction cost in purchasing. Forrester Research calculates that in the old economy, it costs four times the purchase price to process a typical office product requisition, taking as long as one to two weeks to complete the transaction. Forrester also projects that the potential efficiency improvement in business-to-business requisitions will reap between 18 and 45 percent cost savings by employing e-business network connections, with quicker processing, fewer errors, better information, and speedier delivery.

But to balance the potential cost savings, there most definitely will be an offset in slimmer margins. There is simply less profit in doing competitive business over the Internet. It brings greater opportunity but requires greater efficiency. And the one area where the greatest payback can be achieved without additional capital expenditure for capacity expansion is in boosting throughput on the line in the factory.

We believe the possibilities for the future are unrestricted—for those capable of the truth of this basic axiom: What you learn along the way *does* matter, but you must be alert to constantly refresh the context in which it has meaning. This is as true for our customers, the people involved in adding value to raw materials in order to meet the needs of their customers, as it is for Lilly Software and every other software vendor who wants to stay in the market. For us, it has been a lesson hard learned, as are most lessons of important sustaining value. But for us, this axiom goes to the heart of our business; it is instilled in the principles of the company, with our greatest challenge being to not loose sight of these principles as we move forward, to not risk the temptation of complacency to merely get by.

Looking back, the story of the value of information technology has been driven primarily by advances made in computer hardware technology. In contrast, software application engineering has only recently begun to show promise of catching up with the pace of development in hardware as we enter a new millennium filled with promise. Ironically, software design, especially the design of manufacturing applications, was initially held in check by constraints in hardware memory and processing speed. But resistance to change, measured out in an unwillingness to question first-order assumptions within the software applications industry, has also retarded advancement.

While gains have been made in manufacturing control system design, the implications of the history of application development continue to have real significance to manufacturers even as change seems well seated to prosper in the days ahead. Time has taught us that we all must continue to actively question assumptions we are temped to take for granted. This includes manufacturers looking to deploy information in a world where change always threatens to undercut the foundations of their competitiveness. This is nowhere more true than it is for CEOs and senior level managers.

I would offer seven guiding principles for anyone with the charter of bottom line accountability for the health of their manufacturing enterprise. Some of these points seem to beg the obvious, but from my experience, it is often the obvious of which we are most oblivious. I offer these points respectful of the myriad conflicting demands every CEO and senior-level manager faces. But in today's world, if you are in manufacturing, there is nothing so critically demanding of your attention as what goes on inside your plant, out on the shop floor.

1. **Become personally committed to the due diligence involved in the selection of manufacturing control system technology for managing production.** Technology is a sophisticated financial investment, but it should not be left to your financial department nor to your data processing department to solely determine the priority of requirements in the selection of technology at the heart and soul of your existence: your ability to produce to the demands of the market.
2. **Become knowledgeable of the legacy—positive and negative—of information technology, so that you can make informed decisions free of feelings of intimidation.** Increasingly, companies across a broad spectrum of industries are turning to various technologies as critical tools for improving strategic competitiveness. Recognize the opportunity and risk that this implies and respond accordingly.
3. **Demand that technology vendors explain functionality and value without the obfuscation that carries the implicit caveat "trust me."** If it does not make sense, trust your own instinct that something is not right. If they cannot explain their product in simple terms, it is their problem, not yours. If you, the CEO or plant manager, do not understand it, do not buy it. It will not work.
4. **On a tactical level—with grave strategic importance—avoid level-by-level bills, standard leadtimes, and infinite scheduling as valid in the running of your business.** As surprising as it may seem, these are still practices with a real of approval in the industry and integral to many MCS packages on the market today. If you currently have a system that employs these principles, get rid of it before it runs your operations into the ground. If anyone comes bearing gifts in these guises, turn them away. These are cancers that will erode the competitiveness of your company.
5. **Be hands-on involved in the details of your company's ability to produce: to meet schedules, keep commitments to customers, and increase profitability.** If you are a manufacturer, there is nothing more critical to the health of your company than meeting schedules, keeping commitments to your customers, and increasing profitability. Common Sense Throughput provides a disciplined means of meeting all three of these objectives simultaneously.
6. **Manage constraints on throughput as the premiere leverage point in your plant for increasing sustainable profitability.** Improved

throughput without increased capital cost puts more money on the bottom line faster than any other means available.

7. **Question assumptions—your own as well as those layered beneath the apparent sound reasoning of others.** If the goal is to be creative in the solutions we apply, and if the solutions we devise are dependent upon the questions we ask, *and* if the questions we ask are dependent upon the assumptions we hold, doesn't it make sense to first articulate and then question our assumptions? As difficult as this is—and it is no small challenge to uncover the protective belt of assumptions that frame our worldview—this is where the greatest opportunity exists for breakthrough results. History abounds with examples of this. So be bold. Write your own history. Question assumptions.

Appendix A: Case Studies

VLOC

Profile

Company:	VLOC
Parent Co:	II-VI Incorporated
U.S. Location:	Port Richey, Florida
U.S. Employees:	120
Manufacturing Type:	Multiple batch process
Products:	Optics, coatings and crystals for near-infrared laser markets

Founded as Virgo Optics in 1979, VLOC has achieved a remarkably diverse product line, accelerated revenues, and heightened market visibility through a series of strategic acquisitions and mergers.

Today, VLOC is a key division of II-VT Incorporated based in Saxonburg, PA, a publicly-held company that specializes in the manufacturing of infrared optical components and materials for high-powered industrial lasers and military sensing systems. VLOC is one of a few companies in the world that grows, or fabricates, crystals for the burgeoning laser industry. Combined, VLOC and II-VT address the complete spectrum of needs for the global laser components market.

VLOC's uniqueness stems from a diversity of product coupled with the vertical integration of the manufacturing process. Production consists of a number of distinct operations for the fabrication of the company's products, more than 90 percent of which are custom-engineered. It is a highly specialized environment where precision, quality, and timely delivery are of utmost importance to its customers, and to VLOC's ongoing success.

VLOC's corporate motto sums up its dedication to quality: "Do It Right The First Time, On Time, Every Time."

The Challenge

VLOC used proprietary hardware and software for years, but in 1995, its parent corporation gave the company a clear directive to find a software package dedicated to manufacturing in order to manage its demanding production environment and ensuing growth. VLOC specifically wanted a solution with a Windows interface that could satisfy its growth needs for the next five years. The company was also looking for a user-friendly product that did not require customization or extensive training.

The Solution

After much research, VLOC selected and implemented its new Enterprise Resource Planning (ERP) system across three facilities in two locations. VLOC has an intense monthly business cycle that must be closely managed at all times and needed to consolidate financials across all facilities. Integration was also important because the company uses numerous dial-in capabilities and remote sales personnel must be able to receive accurate, up-to-date information when they place, check, and track orders.

With instant access to information, VLOC has streamlined its business processes and improved its ability to deliver on time. Daily production meetings rely on reports from the system. With a common database, the system automatically updates any changes, so managers know the correct status of every job and can make more profitable, knowledgeable decisions.

VLOC has significantly improved customer service by using tools that provide accurate forecasts and delivery dates. With online inventory, sales personnel can give more reliable quotes and forecasts. For the first time in the company's history, it maintains an accurate picture of inventory as it fluctuates within the monthly business cycle. Since VLOC became part of a public company, this has been an important point of cost control.

With its ERP system in place, VLOC has the tools to ensure the continual growth and profitability of the company, as well as to maintain its standards for quality products, on-time deliveries, and customer satisfaction.

Diagnostic Ultrasound

Profile

Company:	Diagnostic Ultrasound Corporation
U.S. Location:	Redmond, Washington
Employees:	100 Worldwide
Manufacturing Type:	Make-to-stock
Products:	Medical

Gerald McMorrow founded Diagnostic Ultrasound Corporation in 1984 as a classic, one-man engineering start-up. McMorrow developed a proprietary ultrasound Doppler device akin to the stethoscope that enables physicians to listen to and characterize vascular blood flow. This initial product offering was the company's mainstay for three to four years and is still marketed by Diagnostic Ultrasound's direct sales force and network of worldwide distributors.

In 1989, the company made a strategic acquisition of the rights to what has since become its premier product offering: Bladderscan, a portable ultrasound scanner that permits noninvasive measurement of bladder volume. Bladderscan and its derivative products are unique for bringing noninvasive technology to the patient at a fraction of the cost, inconvenience, and risk of infection associated with conventional diagnostic methodologies such as catheterization. A sophisticated blend of hardware and software makes Diagnostic Ultrasound devices easy to use, with little training required. The product line consists of several types of devices suitable for both home use and a variety of clinical settings.

In recent years, the company has undergone steady, double-digit growth and, having achieved IS09000 certification, is now poised for rapid expansion of its global markets. To date, Bladderscan technology has no known competition and, thus, affords Diagnostic Ultrasound a unique opportunity in the worldwide market for medical instrumentation.

The Challenge

The acquisition of Bladderscan technology proved to be a turning point for Diagnostic Ultrasound, propelling the company into the global arena and requiring a shift from its original engineer-to-order environment to the highly regulated, make-to-stock production mode that characterizes its business today.

As Diagnostic Ultrasound began this shift, its business technology requirements also changed. Supported by nonintegrated, DOS-based proprietary applications, management had no common frame of reference for assessing mission-critical operations within the business. The data was there, amid multiple separate interfaces, but the applications did not integrate, making the integrity of the data questionable.

One of the most difficult problems the company faced was the absence of a real-time application to manage the growing manufacturing environment, which today includes production of all components—hardware, software, electronic circuit boards—of the equipment, everything except the plastic molds which encase the units. The company needed a manufacturing application with integrated financials that would support the information requirements for the company worldwide across its subsidiaries.

Within the make-to-stock environment, Diagnostic Ultrasound's growth created difficulties in managing cost effective material availability. It had trouble scheduling production and maintaining the right mix of materials in inventory, which resulted in difficulties delivering on time to its distributors.

Although the company had already migrated to Windows, it still lacked an integrated database to help streamline operations and facilitate key decision making. Too many spreadsheets and not enough people made managing the business difficult. Diagnostic Ultrasound was looking for a system that could support its growth for at least the next five to ten years.

The Solution

After implementing an integrated Windows-based Enterprise Resource Planning (ERP) system, the company could consolidate financials across its three subsidiary sites. Almost immediately the company could forecast material requirements with greater accuracy, schedule production with greater efficiency, and meet the delivery dates promised to its distributors.

With advanced scheduling tools, Diagnostic Ultrasound could finally look ahead and plan its production according to changing customer demands. By having all of its information available at the click of a button, the company could now eliminate paper trails and manage specific, uniform FDA and IS09000 requirements. For example, using lot serial traceability, the company could more accurately track every unit produced.

Today, with material requirements planning assured, production streamlined, accounting functions integrated and online, and a product without competition ready for multiple and eager markets, Diagnostic Ultrasound is well positioned for the future.

Godwin–SBO

Profile

Company:	Godwin–SBO, L.P.
Parent Co:	Schoeller–Bleckmann Oilfield Equipment AG , Austria
U.S. Location:	Houston, Texas
Employees:	91
Manufacturing Type:	Make-to-order
Products:	Components for energy equipment industry Precision machined parts for directional drilling

Godwin–SBO, L.P., formerly Godwin Machine Works, builds components primarily for the energy equipment industry, specializing in parts for directional drilling. The company performs most of its production on computerized numerical control (CNC) machinery with the remaining production being done on conventional equipment.

Tight control is hard-won at Godwin–SBO. At any one time, the company has more than 1000 custom jobs in process. Due dates are as close as a day and as far off as six months. Tolerances are as low as 0.0001 inch. And every morning, production managers arrive at Godwin to find new orders from such energy-industry giants as Halliburton, Baker Hughes, and Schlumberger. Godwin has nearly doubled its personnel since 1995, from 50 to 91 employees, to keep up with the volume of orders.

The Challenge

Until 1995, Godwin operated successfully as a 30 to 50 man shop. Production managers would receive a stack of work, file it by due date, and rely on their knowledge of how long a particular job would take to run for scheduling. As Godwin's reputation and customer base grew, scheduling became difficult especially with labor, machine, and material availability variables. As its paper-based system became ineffective, the company missed delivery deadlines and customer satisfaction began to decrease.

At the time, Godwin ran two different software systems: a manufacturing package and an accounting package. Both applications ran on different platforms and contained over 200 individual database files, often with duplicate information. While the accounting package was satisfactory, the production scheduler could not adequately track jobs in progress or shop resource utilization. Godwin became days, even weeks behind for given jobs. Wasted time affected production as employees continuously set up and broke down machines to meet partial orders, and administrative personnel repetitively entered the same data into the two systems.

The Solution

Godwin needed a system with tighter scheduling capabilities—a system that could integrate the 200 individual database files from accounting and production into a single database.

After the company implemented an Enterprise Resource Planning (ERP) solution with advanced planning and scheduling capabilities, Godwin employees worked the same number of hours but production levels skyrocketed and sales doubled in two years. Godwin could finally manipulate the production schedule until all jobs fit into a realistic timeframe.

Today, Godwin department managers consistently use the integrated system to foresee and prevent production bottlenecks. After a job is in progress, employees can view, in advance, the impact of any scheduling changes on the final delivery date. At any time, if a machine is down, an operator is sick, or there is just an abundance of work for a resource, Godwin can determine exactly how that will affect a delivery date.

Godwin's customer base is delighted with the nearly 100 percent on-time delivery and with the quality service Godwin can now provide. When customers call for a status or a quote, they can have that information immediately, rather than waiting for the customer service representative to search for the answer and get back to them.

After installing the integrated solution, Godwin has added only two administrative people, but has increased its production staff more than 25 percent to handle the larger volume. By maintaining all the data from production to accounting in one database, the company has eliminated redundant data entry and ensured accuracy and consistency. Godwin now controls its shops and workloads tightly and efficiently, and the company continues to expand its facilities and production capacity.

Xomox Corporation

Profile

Company:	Xomox Corporation
Location:	Chihuauha S.A, De C.V., Mexico
U.S. Location:	Cincinnati, Ohio
Parent Company:	Emerson Electric
Employees:	140
Manufacturing Type:	Maquilajora (custom assembly plant)
Products:	Process control valves

Located in the heart of Chihuauha, Mexico, Xomox Corporation has been producing top-of-the-line process control valves since 1956. These valves, embedded in equipment throughout the world, provide improved in-line sealing and superior fugitive emissions control. The superiority of Xomox's valves continues to attract an extensive and esteemed client base, including Celanese Mexico, Pevex, and Procter & Gamble.

The Challenge

Xomox's premier product and respected client list did not result in overall profitability for the company. In fact, Xomox consistently found itself in the red at the end of each quarter, even when it had predicted profit. In an effort to arrest further financial losses, Xomox assigned the problem to a trouble-shooting team.

The team was primarily disappointed with Xomox's spotty record of on-time delivery. Tracing missed delivery dates back to resource availability, the team found that Xomox failed to anticipate material needs prior to production. Materials were, therefore, unavailable for jobs at production time, leading inevitably to missed delivery dates. Xomox Corporation needed a solution which would provide tighter control of materials, improving delivery dates and increasing profit. Xomox also needed an accurate job costing method to create realistic projections, so the company could avoid incorrect decisions based on the theoretical profit of the valves.

The Solution

Upon examining the troubleshooting team's findings, Xomox's parent company, Emerson Electric, recommended that Xomox install an Enterprise Resource Planning (ERP) system to manage and maintain all materials and resources.

After implementing the new ERP system, employees were able to find immediate answers about material availability and could provide customers with fast, accurate delivery and cost estimates. Xomox currently has a 90 percent success rate on meeting delivery dates with backlogs of less than a week.

On-time delivery was just one improvement that Xomox experienced after system implementation. Xomox began creating weekly customized reports about production information and capacity utilization, which allowed the company to see where bottlenecks and incidences of contention occurred. Using these reports, managers realized Xomox's profits were being wasted on excess inventory and poor resource utilization. In fact, Xomox calculated resource utilization was a low 40 percent.

With enhanced reporting and scheduling capabilities, Xomox reduced inventory costs to $2 million from a previous high of $3.5 million, WIP backlog was reduced from 150 per cell to 16 per cell, and aging decreased from 30 to 60 days to 7 days. With these drastic improvements, Xomox has been able to maintain 97 percent inventory reliability, while increasing their resource utilization from 40 to 80 percent.

Most importantly, the company has achieved consistent profitability each and every quarter and can now measure and control its processes. With reports, real-time information, and "what-if" scheduling, Xomox can determine the impact of production decisions and create schedules that consistently increase throughput and profit.

Barudan America, Inc.

Profile

Company:	Barudan America, Inc.
Parent Co:	Barudan Ltd. Japan
U.S. Location:	Solon, Ohio
Employees:	70
Manufacturing Type:	Capital equipment
Products:	Embroidery machines

Barudan America is a leading supplier of single and multihead embroidery machines for small home-based and large industrial users throughout North, South, and Central America. Other divisions around the world support the penetration of Barudan's product line into the global embroidery machine marketplace.

With cultural and manufacturing roots in both America and Japan, the company owes its longevity to quality, innovation, and the ability to manufacture machines that use modern technology to recreate the ancient art of embroidery. The fashion industry's increasing use of embroidery to brand retail goods—from baseball caps to T-shirts to tote bags—has placed in demand computer driven embroidery equipment, such as that manufactured by Barudan.

The Challenge

Prior to its acquisition by Barudan Ltd. in 1985, the company manufactured single-head machines used primarily for monogramming. As Barudan America, the company's product line expanded to include multihead machines. This resulted in increased sales and production, as well as an explosion in the number of bills of material.

In 1992, with the help of an outside consultant, Barudan began to establish criteria for a software solution to support the entire business. Among other features, it needed a product with a Windows interface and financials (with multicurrency capability) integrated with manufacturing as the principle application.

The Solution

The single, most important benefit derived almost immediately from Barudan's rapid implementation of its new Enterprise Resource Planning (ERP) system was in the area of costing. The switch from manually calculating costs on a quarterly basis to real-time reporting has kept Barudan current with actual cost of goods sold. Now the company receives up-to-date costing on every item in stock, as well as the latest pricing from suppliers, for each new quote or order. Barudan eliminated the time-consuming days of calculating ineffective cost schedules. The company has also made aggregate quarterly discrepancies and the accompanying year-end write offs negligible.

Using Material Requirements Planning (MRP) functionality, Barudan always knows its material needs ahead of schedule, which has enabled the company to effectively implement Just-In-Time inventory practice and assign an accurate value to its in-process inventory, including labor. Management can produce more accurate reports of forecasted sales because they know costs and can predict secure delivery dates. The company can easily measure its overall financial health at any time and use this information to satisfy its lending partners.

The new ERP system has helped the company manage its production environment and business flow. Barudan has successfully reengineered its business to meet ever changing market demands and stay competitive. With costs under control and the data necessary to make informed pricing decisions, Barudan has the systems and technology in place to support its business for the future.

B Appendix B: Time Line

Late 1800s

Rapid expansion of industrial productivity, transportation, communications, and interstate business as a result of the blossoming of the industrial revolution leads to "crisis of control" of transactions, resulting in the growth of the mechanical office equipment industry (keypunches, sorters, tabulators).

1890: First commercial application of punch card processing developed by Herman Hollerith (1860-1929) slashed census tabulation from the typical ten-year cycle to two years.

1900s

Railroads are the leading industry to widely adopt new mechanical office equipment.

1910s

The First World War defines the trend of the century of a world growing smaller and more interconnected.

1911: Consolidation of several small companies into the Computing-Tabulating-Recording (CTR) Company (later to become International Business Machines).

1913: Henry Ford begins producing automobiles on an "assembly line."

1914: Thomas J. Watson, Sr. (1877-1956) becomes general manager of CTR.

The assassination of the Archduke Francis Ferdinand, heir to the Austrian Empire, in Sarajevo precipitates the beginning of World War I.

1917: After struggling to remain neutral, the United States enters the war in Europe.

1918: The collapse of the German army brings an end to the First World War.

1920s

Statistical theory is formulated. This would later evolve into statistical quality control (SQC), promulgated by W. Edwards Deming and Joseph Juran in the 1950s and 1960s.

Manufacturing cost accounting is developed; 80 percent of all costs are associated to direct labor.

1920: The 19th Amendment to the Constitution is ratified, granting women the right to vote.

1924: CTR becomes International Business Machines (IBM).

1928: IBM introduces the 80-column punch card, employing rectangular holes for better space utilization of wire brushes used for "reading." Watson markets as the "IBM card" and wins critical market acceptance over Remington Rand's 90-column card.

1929: Stock market crashes, bringing on the Great Depression.

1930s

Bell Laboratories begins to systematically apply SQC principles, inaugurating the shift from *inspection* to *elimination* of defects.

The Great Depression grips the nation and the world for much of the decade, being ultimately foreshortened by the launch of the war efforts of the 1940s.

1935: IBM introduces the 600 Series punch card-driven multiplying machine, a relay-based arithmetic unit capable of multiplying two numbers in 1 second.

President Franklin D. Roosevelt promotes passage of the Social Security Act in response to the Depression, creating instant market demand for information processing equipment. IBM's annual revenues, which had stagnated below $20 million, jumps to $38 million.

1937: Konrad Zuse (1910–1995) develops binary memory model from "first principles."

1939: The German invasion of Poland draws declarations of war from Great Britain and France, thus beginning World War II.

1940s

The United States Justice Department investigates National Cash Register (NCR) for violation of the Sherman Antitrust statutes.

Rising cost of labor and increases in government paperwork spur demand for increased computational power.

1941: The Japanese attack Pearl Harbor, causing the United States to formally enter World War II.

1943: Drs. J. Presper Eckert and John W. Mauchly develop a calculating device for computing ballistic firing tables for munitions.

1945: The charter of the United Nations is drafted in San Francisco.

Germany surrenders unconditionally, ending World War II in Europe.

The United States drops atomic bombs on two Japanese cities, Hiroshima and Nagasaki.

The Japanese surrender, ending World War II.

1946: Eckert and Mauchly create the Electronic Numerical Integrator Computer (ENIAC), the world's first electronic digital computer. ENIAC employs 18,000 vacuum tubes and fills a room 30 x 50 feet. It was used after the war at Los Alamos Scientific Laboratories in New Mexico to develop the first hydrogen bomb.

Atomic Energy Commission is created to promote peaceful application of atomic power.

Winston Churchill coins the term the "Iron Curtain" in reference to the Soviet bloc nations of eastern Europe. The Cold War begins.

The Simmons Company of Petersburg, Virginia offers an "electric blanket" for $39.95.

1947: B. F. Goodrich develops the "tubeless" tire.

1948: The IBM 604 (in the 600 Series) is capable of being "programmed" by using two plugboard panels and performs 60 operations.

The transistor is invented at Bell Labs.

Columbia Records introduces the 33 1/3 "long playing" record at the Waldorf Astoria Hotel in New York City.

Late 1940s: American and British engineers devise three viable methods for computer memory: magnetic drum, acoustic delay time, and cathode ray tube.

1950s

The United States Justice Department investigates IBM for violations of the Sherman Antitrust statues for its dominance in the punch card equipment market. IBM signs a consent decree that it will sell equipment in addition to its long-standing practice of leasing.

Customer demand that was put on hold during the Depression of the 1930s and the War Years of the 1940s is unleashed.

America starts construction of the Interstate Highway System under appropriations for future war preparedness.

1950: Eckert-Mauchly's company is acquired by Remington Rand, who begins commercial development of the UNIVAC computer, funded, in part, by the U. S. Census Bureau and Prudential Insurance Company.

Senator Joseph McCarthy announces in Wheeling, West Virginia, "I have here a list ..." leading eventually to the Army-McCarthy Congressional investigations into membership of the Communist party.

United States military troops cross the 38th parallel, dividing North and South Korea.

1951: The Philadelphia branch of the U. S. Census takes delivery of the first UNIVAC computer, weighing eight tons, with 5,000 vacuum tubes, and capable of 1,000 calculations a second, inaugurating commercial computing. Price: $159,000.

President Harry Truman relieves General Douglas McArthur of his command for suggesting the solution to the Korean conflict is to attack Communist China.

Julius and Ethel Rosenberg are sentenced to death after being convicted of conspiring to steal atomic secrets.

Remington Rand's UNIVAC I uses teletype keyboard and printer for input/output devices.

1952: General Dwight D. Eisenhower is elected President of the United States over Adlai Stevenson.

Vice President Richard Nixon gives his "Checkers" speech, confirming that he accepted over $18,000 from supporters, but denies any impropriety. He says he will keep a dog named Checkers that was also given to him by supporters.

1953: Charles E. Wilson, president of General Motors, is named Secretary of Defense, but is compelled to sell his stock in the company. "I thought what was good for the country," he says at the time, "was good for General Motors, and vice versa."

The first magnetic core memory system goes into MIT's Whirlwind computer.

IBM launches development of the 702 Series computer for the commercial market.

IBM deploys the 650 Magnetic Drum calculator.

GE Appliance Park in Louisville, Kentucky, installs a payroll system, the first commercial computer application.

The first hydrogen bomb is tested.

Army-McCarthy Congressional hearings are formally launched.

The landmark U. S. Supreme Court case, Brown vs. Board of Education, declares that "separate but equal" segregated schools are unconstitutional.

The FORTRAN computer programming language is developed for scientific and engineering applications.

IBM announces the 650, the first mass-produced computer.

1955: The United States gives South Vietnam its first aid package, totaling $216 million.

IBM surpasses Remington Rand for the first time in the number of installed computers.

In the United States, 96 percent of all radios sold are manufactured domestically.

The Salk polio vaccine is made widely available for the first time.

1957: IBM introduces the RAMAC 305 random access disk storage system.

Russia launches "Sputnik," the first artificial earth satellite.

Nine Afro-American students integrate Little Rock High School in Arkansas with the assistance of the U. S. Army.

Digital Equipment Corporation (DEC) and Control Data Corporation are founded.

Honeywell unveils its Datamatic 1,000 computer.

The first FORTRAN compiler is connected to an IBM 704 computer.

The American Production and Inventory Control Society (APICS) is formed.

Explorer 1 becomes the first successful American satellite to be launched into space.

The National Aeronautics and Space Administration (NASA) is created.

The first regularly scheduled trans-Atlantic jet passenger service from New York to London commences with a flight time of six hours, twelve minutes.

The "beatnik" movement spreads from San Francisco across the United States and Europe.

1959: Jack Kirby of Texas Instruments and Robert Noyce of Fairchild Semiconductor file separate patents for the first integrated circuit.

Fidel Castro and his band of revolutionaries march on Havana.

The Common Business Oriented Language (COBOL) is formulated, becoming the first computer programming language to use simple English terms.

IBM announces the 1401 computer.

Hitachi, NEC, and Oki Electronics debut computers in the Japanese market.

Pioneer 4 flies past the moon on its way to orbit the sun.

NASA names the first seven astronauts of the Mercury Program.

Alaska becomes the 49th state.

1960s

The United States Justice Department investigates IBM for violations of the Sherman Antitrust statues for its dominance in the computer market. Atmospheric testing of nuclear weapons motivates many Americans to build underground "fallout shelters" in their backyards.

American manufacturing rules supreme in the world.

1960: Digital Computer Corporation (DEC) introduces the PDP-1 minicomputer with CRT and keyboard, inspiring MIT students to write the first computer game.

IBM employee count tops 100,000.

An IBM RAMAC 305 computer manages athlete standings at the winter Olympics in Squaw Valley, California.

Hawaii becomes the 50th state.

The nuclear-powered submarine, USS Triton, circumnavigates the globe, traveling almost the entire 41,500 miles of its voyage underwater in 84 days.

A Polaris missile is launched from a submarine for the first time.

A host of new computers is launched by various companies, including CDC, Sperry-Rand, General Precision, and Philco.

IBM brings to market its 1401 computer, based on solid state electronics, providing a relatively low cost system that sparks big growth for the company.

Time-phased material requirements planning, what would formally become MRP, begins in earnest.

Sony introduces its first 8-inch miniaturized monochrome television.

John F. Kennedy is elected President of the United States, defeating Richard M. Nixon.

Dick Lilly joins IBM.

1961: IBM delivers its first STRETCH "super computer" to the Los Alamos Scientific Laboratory of the U. S. Atomic Energy Commission, the most complex electronic device yet configured, incorporating disk drives capable of a multiple read/write function.

Outgoing President Eisenhower warns of the danger of the developing "military-industrial complex" in his farewell address.

Soviet cosmonaut Yuri Gagarin becomes the first human in space.

Three weeks later, astronaut Alan B. Shepard, Jr., becomes the first American in space, riding atop a Redstone rocket from Cape Canaveral. His flight is tracked by an IBM 7090 computer.

An IBM RAMAC 305 is installed at a New Jersey Volkswagen dealership to be used for processing 20,000 new car orders for the year and 6,000 parts weekly.

The Association of Data Processing Service Organizations (ADAPSO) is formed.

East Germany erects the Berlin Wall to stop its citizens from fleeing to the West.

1962: The Cuban Missile Crisis brings the United States and Russia, the two superpowers, to the threshold of nuclear war. Russia agrees to remove its missiles from Cuba.

Astronaut John Glenn becomes the first American to orbit the earth.

Telstar, the first privately owned satellite (AT&T) is deployed to transmit television programs across the Atlantic.

Mariner 2 transmits data from Venus across 36,000 miles of space.

IBM introduces the first disk file storage system.

APICS membership grows to 2,300.

1963: Sketchpad, the first WYSIWYG interactive drawing tool, is developed as part of an MIT thesis.

IBM gathers its brain trust in White Plains, New York, to begin the design of the manufacturing Production Information and Control System (PICS) in preparation for the launch of a new "family" of computers.

The British rock group, The Beatles, becomes an international phenomenon.

First-class postage jumps from 4 to 5 cents.

President Kennedy is assassinated.

ASCII is adopted as a common file format.

GE releases IDS, the first database management system.

The first CAD-designed parts are built at GM.

American Airlines' SABRE computerized reservation system goes online.

1964: DEC announces the PDP-8 minicomputer, the first computer in a cabinet. Price: $16,200.

IBM announces the System 360 family of seven computers, initiating the concept of investment protection for companies needing to migrate to larger systems. The 360, which deploys an operating system, creates a huge demand for application programs.

The United States launches a Saturn booster rocket with a 10-ton payload, marking the first time the U. S. has launched an object heavier than one launched by the Soviet Union.

Cassius Clay, later to be known as Muhammad Ali, defeats Sonny Liston to claim boxing's world heavyweight champion title.

The United State's Ranger 7 spacecraft crash lands on the moon after transmitting 4,000 close-ups of the lunar surface.

Martin Luther King is awarded the Nobel Peace Prize.

Ted Nelson coins the terms *hypertext* and *hypermedia.*

1965: The first American troops, 3,500 Marines, arrive in Vietnam.

Mariner 4 reaches Mars after an eight-month journey and begins transmitting pictures of the red planet.

President Lyndon Johnson launches the aerial bombing of North Vietnam.

APICS membership grows to 5,000.

1966: The hallucinogenic LSD and mood altering marijuana become a growing national concern.

The United States commits over 190,000 ground troops to Vietnam.

President Johnson approves bombing of Hanoi.

Orbiter 1 transmits the first photographs of earth as seen from the moon's surface.

The percentage of American-made radios sold in the U.S. that are made here drops to 30 percent.

1967: American youth flock to San Francisco's Haight Ashbury district for the "summer of love."

Race riots erupt in over 120 U. S. cities.

Virgil Grisson, Edward White, and Roger Chaffee become the first American astronauts to perish in the space program, dying in a fire aboard an Apollo spacecraft at Cape Canaveral.

Sixty-three nations sign a space treaty prohibiting orbiting nuclear weapons and any claim to moons or planets.

1968: Martin Luther King and Robert Kennedy are assassinated.

Alabama Governor George Wallace runs for President, but is paralyzed by an assassin's bullet.

President Johnson announces he will not run for President in the upcoming elections.

Richard Nixon is elected President.

Intel Corporation is founded.

The first software patent for a "sort" routine is granted to Martin Goetz.

The first mouse interface is demonstrated.

APICS hires its first executive director.

Dick Lilly leaves IBM to create Software International.

1969: Astronaut Neil Armstrong becomes the first human to set foot on the moon.

IBM announces the "unbundling" of software and services from hardware sales, formally creating the independent software industry.

Woodstock, New York, a tiny farming community, draws 300,000 people attend a four-day music festival.

A nationwide Vietnam Moratorium Day is held, protesting the war in Vietnam.

Work on the ARPANET packet switching network is started.

The UNIX operating system is created by Bell Labs.

1970s

The United States Justice Department begins an investigation of AT&T into allegations it has violated the Sherman Antitrust statues. The investigation will remain ongoing into the 1980s, when the Justice Department orders the break up of AT&T.

The emerging trend of offshore competitors gaining and strengthening their footholds in the American market based on quality and cost factors continues to gather force. Growing inflation at home makes it difficult for manufacturers to invest in capital improvements to stay competitive.

1970: The first Earth Day is held.

Amdahl Computers is formed, offering IBM plug-compatible mainframe computers.

The first four nodes of ARPANET are connected, establishing what will grow into the Internet.

APICS membership grows to 8,000.

1971: The microprocessor is invented by Ted Hoff and enters the market as the Intel 4004.

IBM introduces the 3270 mainframe terminal, establishing its character-based terminal as an industry standard.

ICP presents its first "Million Dollar Awards" to 24 software companies.

1972: Seymour Cray founds Cray Computers to develop and commercially market "super computers."

The first graphical user interface (GUI) is created using Xerox PARC's Smalltalk programming environment.

APICS launches the "MRP Crusade."

The DOW passes 1,000 for the first time.

President Nixon visits China.

The last U. S. ground troops are withdrawn from Vietnam.

Arab terrorists kill 11 Israeli athletes in the Olympic Village in Munich, Germany.

The Democratic Campaign office in the Watergate hotel is broken into.

1973: APICS certification program begins Certification in Production and Inventory Management (CPIM), and the offering of its first two courses in its educational program: inventory planning and forecasting.

Skylab space station, an 86-ton structure, goes into orbit.

Ethernet is invented at Xerox PARC by Robert Metcalfe.

Formal peace accords for the war in Vietnam are signed in Paris.

Oliver Wight begins formal evaluation of commercial MRP software packages.

1974: The Altair 8800 microcomputer processor is introduced, based on Intel's 8080 integrated circuit.

The Skylab space crew returns after 84 days and a 34.5 million-mile journey, circling the earth 1,214 times.

IBM announces its System Network Architecture (SNA).

IBM designs the first RISC computer.

Wang Corporation delivers its word processing system to the market.

The Middle-East Oil Embargo cuts oil shipments and boosts U. S. gas prices.

Hank Aaron breaks Babe Ruth's lifetime homerun record with 715.

Nixon resigns.

President Ford pardons Nixon of any federal offenses he might have committed.

Dick Lilly leaves Software International.

1975: APICS membership grows to 12,000.

Production of radios in the United States approaches zero.

Joe Orlicky publishes his book *Material Requirements Planning*, beginning true formalization of the MRP discipline.

Unemployment stands at 6.5 million, the highest in 13 years.

United States Embassy personnel and pro-American Vietnamese nationals evaculate Saigon by helicopter, ending the United States' presence in Vietnam.

Steve Wozniak, a high school student (later to cofound Apple Computers with his classmate, Steve Jobs) declares to his father that instead of buying a house when he grows up, he will save his money to buy a computer and "become the one person who owns a computer in the world."

1976: Apple Computer Corporation is created.

Microsoft Corporation is created.

1977: "Hobby" computing arrives with the Radio Shack TRS 80 and MIT's Altair personal computers.

Apple II, with integrated keyboard, 16-color graphics, and command line disk operating system, comes to market.

Dick Lilly relocates to Marathon, Florida.

1978: Steve Jobs proposes a next-generation business computer with GUI, which becomes the Apple Lisa.

VisiCalc, created by Dan Bricklin and Bob Frankstan, a text-based spreadsheet, becomes the first "killer app," effectively sparking the PC market boom.

IBM is a $20 billion company.

DEC is a $2 billion company.

APICS membership grows to 27,000.

John Paul II begins his reign as Pope of the Catholic Church.

Price/performance improvement for computing is estimated to have grown 25 percent every year since 1957.

1979: Oliver Wight's "ABCD" MRP II implementation evaluation standard is created.

Xerox, inventor of the modern copier, has manufacturing costs, product development time, and product development teams twice the size of its Japanese competitors.

The number of American autoworkers peaks at 21 million.

Iranian followers of the Ayatollah Khomeini seize the U. S. Embassy in Teheran and hold 52 Americans hostage.

Dick Lilly starts Key Systems with his son, Mike.

Dick Lilly meets Dave Layne and hires him to do programming for him part-time.

1980s

The 1980s are a time of recriminations for what has gone wrong in American manufacturing, as foreign competitors erode America's market share at home and abroad. Corporate "downsizing" to achieve operations that are "mean and lean" gathers momentum during the last half of the decade.

MRP II acquires something of a bad name in time and is supplanted by a series of banner programs, beginning with JIT in the early 1980s, followed by CIM and World-Class Manufacturing.

1980: APICS membership grows to 46,000.

IBM decides to make a personal computer.

Japan surpasses the Big Three Auto Makers in units of production.

John Lennon is shot and killed outside his apartment in New York City.

The World Health Organization announces that smallpox has been eradicated worldwide.

The Mt. St. Helen's volcano erupts in Washington State, sending debris 12 miles high. The sound of the explosion is heard 200 miles away. Thirty-six people are killed.

1981: The Osborne portable computer hits the market.

The Iranian Hostage Crisis is ended with the release of all hostages.

Compaq Computer Corporation is created.

IBM announces the release of its personal computer.

The Hayes Smartmodem 300 comes to market.

Scientists identify the Acquired Immune Deficiency Syndrome (AIDS).

The French TGV, the world's fastest train, goes into service between Paris and Lyons.

President Reagan is shot and critically wounded, but recovers.

Pope John Paul II is shot and critically wounded in St. Peter's Square, but recovers.

Dick Lilly meets Jesse Jones and begins negotiations to design and build a make-to-order oriented MRP system.

1982: Lotus Corporation is formed and announces the Lotus 1-2-3 spreadsheet for the IBM PC, the first "integrated" PC program.

The IBM PC comes to market. It is based on the Intel 8088 chip, has 64 bytes of RAM, 40K bytes of ROM, a single disk drive, and sells for $3,005. It is a runaway success.

Walt Disney World and EPCOT Center are opened in Florida.

The Radio Shack TRS-80 model 16 costs $4,999.

Dick Lilly breaks with the church of MRP to create the first MTO MRP II system.

1983: *Time* magazine names the computer "Man of the Year."

IBM introduces the System/36.

IBM introduces the PC XT with a hard drive.

Oliver Wight dies of throat cancer.

The maiden launch of the space shuttle Challenger carries Sally Ride, the first American woman in space, into orbit.

Pioneer 10 becomes the first spacecraft to leave the solar system-after an eleven-year flight.

1984: Apple ships the first Macintosh, the first mass-market personal computer.

The United States Justice Department breaks up AT&T.

GM buys EDS for $2.5 billion.

IBM announces the PC AT.

The IBM PC Jr. (the "Peanut") is launched, flounders, and dies.

The American Society of Mechanical Engineers designates the IBM 350 Disk Storage Device as an International Historic Mechanical Engineering Landmark.

The number of new vendors entering the MRP II software field reaches a high point, with 30 start-ups.

A dollar's worth of quality-adjusted computer power has shrunk from $73.60 in 1950, to 5 cents.

Key Systems does $1 million in business.

Dick Lilly relocates Key Systems to Salem, New Hampshire.

1985: Microsoft ships Windows 1.0, with 80 employees on staff.

IBM brings to market its second generation PC, based on the Intel 286 microchip.

Between 1975 and 1985, some 14,000 software firms are created in the United States, lifting the United States share of the world software market from under 33 percent to more than 75 percent.

The space shuttle Challenger explodes shortly after liftoff, killing its crew of seven.

Dick Lilly renames his company ProfitKey International.

1986: MRP vendor market entrants number only a handful.

1987: The United States, once the dominant manufacturer of televisions, is down to one manufacturer, Zenith, who has a 15 percent global market share.

Ford Motor Company rebounds from a loss of nearly $1.5 billion in 1980 to a profit of $4.6 billion.

In less than 10 years, Xerox cuts its manufacturing costs in half, dramatically improves quality, and reverses its decline in market share.

American companies that had moved offshore to take advantage of cheaper labor begin to return as world currencies strengthen against the dollar.

The Dow falls 508 points, losing 23 percent in value on "Black Monday."

Windows 2.0 is released.

1988: IBM introduces the AS/400, designed to replace the System/36/38.

U. S. labor costs drop 30 percent below Germany's and 10 percent below Japan's, a dramatic turnaround from 1985 (when it was nearly 60 percent above Japan).

McDonald's opens 20 restaurants in Moscow.

A terrorist bomb explodes aboard a Pan Am flight over Lockerbie, Scotland, killing 270 people.

1989: Tiananmen Square is occupied by pro-democracy Chinese students, is held for seven weeks, then violently retaken by army tanks. Thousands are believed to be killed.

The Exxon Valdez oil tanker runs aground in Alaska, dumping 11 million gallons of crude oil into the sea, creating the world's worst oil spillage.

An earthquake hits San Francisco, killing 36 people and halting the "battle of the bridge," the World Series between the San Francisco Giants and the Oakland Athletics.

The Berlin Wall is demolished.

Mitsubishi buys Rockefeller Center in New York City.

1990s

The United States Justice Department launches its investigation into Microsoft Corporation for alleged violations of the Sherman Antitrust statutes.

Labor content in manufacturing cost accounting has dropped to under 20 percent, and is heading toward 10 percent.

MRP II morphs itself into ERP, which becomes the banner for just about everything being sold into an "enterprise."

"Globalization," "business process reengineering," and "supply chain management" gather cachet as defining themes for the decade.

American-based manufacturing enterprises regain their stature as the most productive force in the world.

Client-server replaces host-dumb terminal computing as the predominant computing architecture.

1990: APICS membership grows to 60,000.

Saddam Hussein's Iraqi troops invade Kuwait.

Mikhail Gorbachev is awarded the Nobel Peace Prize.

Windows 3.0, the first commercially viable version, is released.

1991: APICS offers Certification in Integrated Resource Management (CIRM).

American and allied forces invade Iraq in Operation Desert Storm, forcing Iraq to withdraw from Kuwait.

Gartner Group declares "MRP evolving to ERP."

ProfitKey brings in a new CEO to replace Dick Lilly.

1992: Apple Newton comes to market.

The Internet Society is founded.

Dick Lilly is officially fired from ProfitKey.

Dick Lilly and Dave Layne begin work on a Windows-based manufacturing control system.

Dick Lilly starts Lilly Software Associates.

Lilly Software sells the first VISUAL Manufacturing software package.

1993: The Mosaic web browser comes to market.

The burning of the Branch Davidian compound in Waco, Texas, is the conclusive end of a standoff with federal agents.

The Pentium-based PC hits the market. The Intel Pentium chip delivers 3 million transistors, capable of executing 100 million instructions per second.

Microsoft unveils the NT operating system.

Nelson Mandela shares the Nobel Peace Prize with F. W. de Klerk.

Lilly Software closes $1.5M in software sales.

1994: Tim Berbers-Lee joins MIT and founds the WWW Consortium.

Nelson Mandela is elected president of South Africa.

1995: Microsoft Windows 95 comes to market.

1996: Lilly Software customer base passes 500.

1997: APICS sponsors a total of 270 local chapters.

Netscape Communications is created.

More than 1.3 million new businesses are started in the United States, but more than half of them fail.

Britain relinquishes its sovereignty over Hong Kong.

Lilly Software begins work on its n-tier next-generation system.

1998: APICS membership grows to 72,000.

Lilly Software is named to *Inc.* magazine's 500 list of fastest growing small companies for the first time.

Lilly Software receives a U. S. patent for finite scheduling methodology.

1999: The countdown to the rolling over of computer clocks for Y2K begins in earnest.

Eleven European countries adopt the Euro, a single currency.

E-Commerce blasts onto the scene, filling the press with optimism for the future, despite the impending arrival of Y2K.

Dot-com madness reaches a fevered pitch.

The Dow crosses 10,000 for the first time.

President Bill Clinton faces impeachment charges in the United States Congress.

Software sales of ERP systems slow dramatically as businesses worldwide focus on getting through Y2K.

Corporations assemble emergency response teams to sit through New Year's Eve on site in the event of catastrophe.

The United States relinquishes its control over the Panama Canal Zone.

2000: The world doesn't end; the new millenium arrives without disaster resulting from Y2K.

References

Armstrong, David J., "Getting Things Done," *Harvard Business Review,* November/December 1985, 42-58.

Brooks, Frederick P., Jr., *The Mythical Man-Month* (Reading: Addison-Wesley, 1995), 266.

Browne, Martin R., "MRP: The First Step Toward World-Class Manufacturing," *Manufacturing Week Magazine,* 7 December 1987, 15.

Casti, John L., *Paradigms Lost: Images of Man in the Mirror of Science* (New York: William Morrow and Company, 1989), on Imre Lakatos, an Hungarian educator and philosopher who developed a model, or methodology, for conducting a "scientific research program" (SRP), assessing the controlling nature of assumptions, 35-36.

Degnan, Christa, "Did You Know the First Computers Wore Dresses?" *Mass High Tech,* 5-11 October 1998, 10.

Dertouzos, Michael L., Lester, Richard K., and Solow, Robert M. *Made in America: Regaining the Productive Edge* (Cambridge: MIT Press, 1989), 23-25.

Farrell, Christopher, "Why the Numbers Miss the Point," *Business Week,* 31 July 1995, 78.

Gilder, George, *Microcosm: The quantum Revolution in Economics and Technology* (New York: Simon & Schuster, 1989), 32.

Ging, Jack, past president of American Production and Inventory Control Society (APICS), in an interview with the author, January 1999.

Goddard, Walter, former president of Oliver Wight Education Associates, in an interview with the author, January 1999.

Hans, Stina, "CIM Starts with MRP II," *Managing Automation,* February 1988, 80.

IDC Newsletter, 1987, 1.

Linton, Calvin D., editor-in-chief, *The Bicentennial Almanac* (Nashville: Thomas Nelson Publishers, Inc., 1975), 384-393.

Lotenschtein, Sergio, "Just-In-Time in the MRP II Environment," *P&IM Review,* February 1986, 28-34.

Loveman, Gary, *An Assessment of the Productivity Impact of IT,* Sloan School of Management/MIT Working Paper on Management in the 1990s, 1988, 1.

Orlicky, Joseph, *Material Requirements Planning* (New York: McGraw-Hill, 1975), 219.

Pugh, Emerson W., *Building IBM: Shaping an Industry and Its Technology* (Cambridge: MIT Press, 1996), 89.

Stoddard, H. G., "Romance of Worcester Industry," *The Worcester Historical Society Publication* (September 1945).

The Oliver Wight Companies, *Company Newsletter,* 1985, 2.

Watson, Tom, Jr., "The Secret of IBM's Success," interviewed by David Brousell, *Datamation,* 15 March 1991, 38-39.

Wight, Oliver, *MRP Software Evaluation* (Essex Junction: Manufacturing Software Systems, Inc., 1983), 2.

Worcester (Massachusetts) Historical Society public exhibits, Summer 1998.

World Class Manufacturing Operating Principles for the 1990s and Beyond, National Center for Manufacturing Sciences, January 1989, 3.

Index

M

N

O

T

U

V

W

X